BUILDING CONSTRUCTION DRAFTING AND DESIGN

John Molnar, P.E.

VAN NOSTRAND REINHOLD COMPANY
New York

Library of Congress Catalog Card Number: 85-26457
ISBN 0-442-26438-0

Manufactured in the United States of America

Published by Van Nostrand Reinhold Company Inc.
115 Fifth Avenue
New York, New York 10003

Van Nostrand Reinhold Company Limited
Molly Millars Lane
Wokingham, Berkshire RG11 2PY, England

Van Nostrand Reinhold
480 La Trobe Street
Melbourne, Victoria 3000, Australia

Macmillan of Canada
Division of Canada Publishing Corporation
164 Commander Boulevard
Agincourt, Ontario M1S 3C7, Canada

16 15 14 13 12 11 10 9 8 7 6 5 4 3 2

Library of Cataloging-in-Publication Data

Molnar, John.
   Building construction drafting and design.

   "May 15, 1985."
   Includes index.
   1. Structural drawing. 2. Structural design.
I. Title.
T355.M65 1986   720′.28′4   85-26457
ISBN 0-442-26438-0

To
Granddaughter Cory
Grandson Jonathan
Give a little love to a child
and you get a great deal back

Preface

Building Construction Drafting and Design was written to provide in one book all the necessary information for anyone who has a need to know how to make architectural and engineering drawings from schematics through working drawings.

The purpose of this book is to bring together, in one handy volume, information usually found in separate and specialized publications. The information is presented in an easy-to-read style and is readily understood by both entry level and experienced persons in the building construction industry. This book may be used for home study for those interested in pursuing a career in drafting and design and as a reference text by those already employed who desire to enhance their knowledge and improve their performance. The book may be used as a supplementary text for trade and vocational school students interested in learning about actual practice methods used in A/E offices. College students and graduates considering careers in building design could benefit from the information contained in this book.

Drafters and designers may work for architects, engineers, contractors, design/build firms, or companies. Whether the work involves making building construction drawings, or reading and using prints for construction purposes, one needs to know how drawings are developed, what information is illustrated, what is noted, how sections are cut and where to find them, when and why schedules are made, and what information is contained in the specifications. The person needs to know what contract documents are and the interrelationship between the specifications and the drawings. While drafters and designers are generally not involved in writing specifications, they need to understand how specifications can be used to advantage, and know the difference between what should be in the specifications and what should be on drawings.

Photographs of actual blackline and blueprint drawings have been included in the illustrations. The author acknowledges the tradeoff between the quality of these reproductions and the intended message of the illustrations. These illustrations were carefully selected from numerous sets of schematic, preliminary and construction drawings to show various types of linework and lettering used by practicing architects, engineers, designers and drafters. The message is in the illustrations and not in the quality of the reproduction.

Anyone interested in a career in the building construction industry should not have to search through numerous sources or depend on associates to learn the practical aspects of the business. In this book, the author hopefully provides all the answers.

The author expresses sincere gratitude to all those individuals, companies, associations and publisher that have provided illustrations, technical information and helped in other ways:

American Institute for Design and
  Drafting (AIDD)
102 N. Elm Place, Suite F
Broken Arrow, OK 74012

American Society of Mechanical
  Engineers (ASME)
345 East 47th Street
New York, NY 10017

Chartpak, A Division of Avery
  Products Corporation
One River Road
Leeds, MA 10153

The Construction Specifications
  Institute (CSI)
601 Madison Street
Alexandria, VA 22314

E. I. Du Pont De Nemours & Co. (Inc.)
Wilmington, DE 19898

Electrical Consultant Magazine,
  A Cleworth Publication
One River Road
Cos Cob, CT 06807

Hunt Manufacturing Company
1405 Locust Street
Philadelphia, PA 19102

Institute of Electrical & Electronic
  Engineers (IEEE)
345 East 47th Street
New York, NY 10017

Koh-I-Noor Rapidograph, Inc.
100 North Street
Bloomsbury, NJ 08804

Mayline Company, Inc.
619 North Commerce Street
Sheboygan, WI 53081

Ozalid Corporation
1000 MacArthur Boulevard
Mahwah, NJ 07430

Sidney Scott Smith, AIA, Architect
P. O. Box 582
Moorestown, NJ 08057

Vemco Corporation
766 South Fair Oaks Avenue
Pasadena, CA 91105

Introduction

In 1984 the top 125 firms in the Commercial-Industrial-Institutional (CII) building construction industry had total fees of about $3.5 billion. This group includes Design/Construct, A/E, and Consulting Engineering firms that used drafters and designers to make working drawings as part of the services rendered for a fee. Then consider the thousands of smaller firms, companies with engineering departments, and the many contractors that erect buildings, and it soon becomes obvious that there are many drafters and designers employed in the building construction industry. Employment opportunities are always open to those who have the necessary drafting skills, are proficient, and can make quality drawings.

The best way to describe what drafters do is through definitions. The following is taken from the U. S. Department of Labor, *Dictionary of Occupational Titles,* Fourth Edition, 1977:

DRAFTER—Prepares clear, complete, and accurate working plans and detail drawings from rough or detailed sketches or notes for engineering or manufacturing purposes, according to specified dimensions: Makes final sketch of proposed drawings, checking dimensions of parts, materials to be used, relation of one part to another, and relations of various parts to the whole structure of project. Makes any adjustments or changes necessary or desired, inks in lines and letters on pencil drawings as required. Exercises manual skill in manipulation of triangle, T-square, and other drafting tools. Draws charts for representation of statistical data. Draws finished designs from sketches. Utilizes knowledge of various machines, engineering practices, mathematics, building materials, and other physical sciences to complete drawings. Classifications are made according to type of drafter as: Drafter—Architectural, Drafter—Electrical.

While this defines all aspects of drafting, whether in building construction or in machine and equipment design, there are certain phrases that clearly describe drafters' work in building construction drafting:

- Prepares clear, complete, and accurate working plans
- Prepares detail drawings from rough or detail sketches
- Checks dimensions of parts and materials
- Makes adjustments and changes
- Inks in lines and letters
- Exercises manual skills in the manipulation of . . . drafting tools
- Utilized knowledge of . . . engineering practices, building materials . . . to complete drawings

This book is written to provide the necessary information required to develop proficiency in drafting.

PART I covers drafting basics. Chapter 1 explains the various drafting tools; Chapter 2 covers drafting skills; and Chapter 3 describes the different types of drafting mediums used.

The previously quoted definition states that "Classifications are made according to the types of drafter . . ." and mentions two (although there are more):

DRAFTER, ARCHITECTURAL—Performs duties of DRAFTER by drawing artistic architectural and structural features of any class of buildings and like structures: Delineates designs and details, using drawing instruments. Confirms compliance with building codes. May specialize in planning architectural details according to structural materials used, as DRAFTER, TILE AND MARBLE.

DRAFTER, STRUCTURAL—Performs duties of DRAFTER by drawing plans for structures employing structural reinforcing steel, concrete, masonry, wood, and other structural materials. Produces plans and details of foundations, building frames, floor and roof framing and other structural elements.

Parts II and III of this book address architectural and engineering drafting and drawing development and explain drafting as practiced in professional offices. This material is presented in sufficient detail to be of value to the entry-level drafter and at the same time serve as a valuable reference to experienced drafters and designers.

The work performed by designers is not quite as simply defined as that of drafters, first of all because architectural designers have completely different responsibilities from engineering designers. An architectural designer is generally a senior professional experienced in building design concepts, whereas the engineering designer is usually a senior nonprofessional working under the direction of an engineer and usually in charge of the drafters on the project. A logical advancement for the proficient, experienced engineering drafter is to the position of engineering designer. What a person is called, "drafter" or "designer," depends to some extent upon the policy and size of the A/E firm. In some small architectural offices where the owner is the principal of the firm, designers are considered as advanced drafters.

Designers are usually required, and often expected, to perform calculations, select materials and equipment, write specifications, and make estimates. For drafters and designers this is not necessarily the end of the road. It is possible for the proficient person who has adequate relevant experience to seek the status and rewards of the professional. Most states permit the qualified person to substitute a required number of years of experience in lieu of formal education to qualify for a written examination for licensure as an architect or engineer. Opportunities for the aspiring drafter and designer are manifold.

For drafters and designers to realize their fondest goals of attaining appropriate status and reaping relevant rewards, they must become thoroughly familiar with all aspects of building construction drafting and design. The entry-level drafter must know what tools and drawing mediums are available, learn how to use them to best advantage, and develop a high degree of proficiency.

With experience, the drafter will become more involved in the process of developing drawings and must become familiar with the different kinds of drawings required for the various phases of drawing development.

A good drawing is one with clear and sharp line work, well selected sections and appropriate details, legible and accurate dimensioning, no nonessential lettering, and is complete, yet simple. A good drawing is one that contains all essential information and nothing more. When too much is drawn, time is wasted, and time is money. When too little is shown, extensive and expensive legal situations may arise during the construction phase. The ability to draw and read drawings is fundamental to the building construction industry.

Since contract documents consist of drawings and specifications, the author has included a section on specifications. It is necessary to understand the relationship between drawings and specifications, the function of specifications, and what information is to be included on drawings or written in specifications.

The intent of this book is to provide in one volume all the information necessary for making quality drawings.

JOHN MOLNAR, P.E.

Contents

PART I
DRAFTING BASICS

Chapter 1
Tools of the Trade

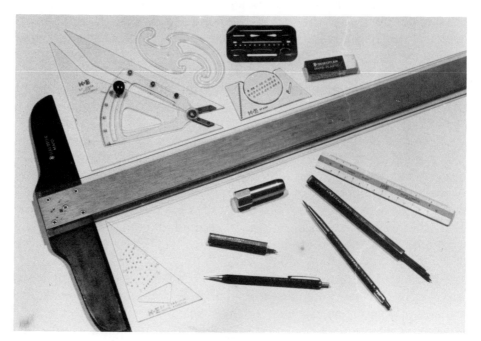

Figure 1-1.

The fundamentals of good drafting must begin with a thorough knowledge of the *tools of the trade.* While this is the foundation on which to build, the super-structure is equally as important. In other words, merely knowing which tools to use is of no value unless appropriate skills in the use of these tools are also developed. The first three chapters discuss the various tools, equipment and materials, drawing mediums and explain how to develop basic skills.

Drafting may be either free-hand or hard line (instrument drawing), using lead or ink on paper or drafting film. Most frequently it may be some combination of the above. Architectural and engineering offices usually dictate the means used to develop drawings.

The serious student of drafting must develop skills in both free-hand and hard-line techniques. Beginners usually start with lead on paper but must be able to work with lead or ink on paper or film.

1. PENCILS AND LEAD

This book will stick with convention using the generic term *lead* in reference to the common *lead pencil* and the *lead* used in lead holders and automatic pencils because it is the accepted practice. Actually, the term *lead pencil* is a misnomer as there is no lead in lead pencils today. Ancient Egyptians and Romans used pencils that were made of lead.

The principal ingredient in most lead pencils and drawing pencils is graphite, a black mineral variety of carbon. Graphite is used because it is safer and makes a much darker mark than lead. Powdered graphite is mixed with pipe clay, and the amount of clay used with the graphite determines the hardness (degree) of the lead. Equal parts of clay and graphite make a hard lead, and as the percentage of graphite to clay is increased, the lead becomes softer.

a. Kinds of Pencils

There are two kinds of pencils: the writing pencil used for general purposes and the drawing pencil used in drafting to develop building construction drawings.

The writing pencil is usually available in only a few different grades (the grade determines the degree of hardness) , i. e., No. 2, No. 2½, No. 3, and No. 4, and they cost less than drawing pencils. While writing pencils may be used for drafting, their use is not recommended for quality work since the lead is often inferior and may scratch or mar the surface of drafting paper. Another difference is the quality of wood used. The wood case of writing pencils is generally made of pine whereas drawing pencil cases are made of red cedar or redwood.

Drawing pencils are manufactured in 17 degrees of uniform, accurately measured, evenly spaced hardness ranging from 6B, soft to 9H, very hard, but may be available in 20 degrees from 8B to 10H. The pencils have imprinted degree markings on the wood case and are generally also color coded for quick identification. These pencil cases are hexagon shaped for comfortable handling and to prevent rolling off sloped drafting tables (Fig. 1-2). The HB grade of drawing pencil compares with a No. 2 writing pencil.

Table 1-1 shows the general usage of the various types of drawing lead pencils (also applies to lead in lead holders) used for drafting. The specific degree of hardness selected depends upon personal preference. For example, a heavy-handed drafter may require a much harder lead to generate the same darkness as a light-handed person. Also, the lead hardness used depends upon the intended line weight: layout lines—light; equipment outlines—heavier; section cuts—broad and dark; pipes, etc.,—darkest.

In addition to the graphite pencils, there are others that need mentioning even though they are not generally used for drafting. They may be found in professional offices and are used for architectural renderings, sketching, checking drawings, and marking up changes on drawings.

Carbon pencils are a greaseless medium, produce a deep black mark, and

Figure 1-2. A number 2H drawing pencil is used for general-purpose drafting and by some drafters for fine line work. (Courtesy of Koh-I-Noor Rapidograph, Inc.)

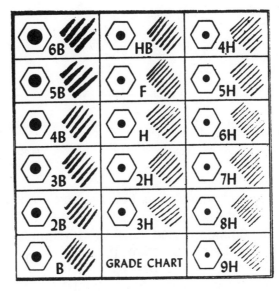

Figure 1-3. Drawing pencil chart indicates quality (darkness) of mark made with the different degrees of hardness.

do not have the shine generally associated with graphite. Used primarily for sketching and free-hand work, they are wood-encased and available in soft (equal in blackness to about 6B), medium (equal to about 4B), and hard (equal to 2B).

A China marker consists of a paper-wrapped, peel-off case with a wax-type crayon lead that produces a very black mark ideal for circling up drawing changes on the underside of polyester film. It may be used on all glossy surfaces, such as film, glass, metal, and china, hence the name.

Plastic-base (polymer) lead pencils have as their base ingredient a special plastic formulation instead of graphite and clay and were created specifically for drawing on polyester films prepared with a matte surface. It is a smudge-proof

Table 1-1. Suggested Lead Selection

GENERAL USE	HARDNESS	SKETCHING	DRAFTING	FINE LINE
	8B	Z		
	7B	Z		
Soft for	6B	Z		
sketching and	5B	Z		
renderings	4B	Z		
	3B	Y		
	2B	Y	Z	
	B	X	Y	Z
	HB	X	Y	Z
	F	Y	X	Y
Medium for	H	Y	X	Y
general-purpose	2H	Y	X	X
drafting	3H	Z	X	X
	4H	Y	X	
Hard for	5H	Z	X	
accuracy and	6H	Z	Y	
fine line work	7H	Z		
	8H	Z		

where: X = First choice
Y = Second choice
Z = Third choice

KOLOR-KODE DESIGNATION

9H Grey	F Light Blue
8H Light Green	HB Yellow
7H Pink	B Purple
6H Silver	2B White
5H Orange	3B Lime
4H Dark Green	4B Light Violet
3H Black	5B Dark Red
2H Light Red	6B Dark Blue
H Brown	

Figure 1-4. Drawing pencil Kolor-Kode designations for quick and easy identification. (Courtesy of Koh-I-Noor Rapidograph, Inc.)

drawing lead with excellent adhesion and erases easily with a good quality soft white vinyl eraser such as FaberCastell Vinyl Magic-Rub. Drawing pencils and leads are available in commonly used degrees of hardness: HB, F, H, 2H, 3H, and 4H, depending upon the manufacturer (Fig. 1-4).

There are numerous other types of pencils for various uses outside the scope of this book but should be mentioned. For example, test scoring pencils sensitive to electric current; colored indelible copying pencils; colored thin-lead pencils for checking, copying, and coloring: lithographic pencils containing a special black graphite lead for writing on master sheets and plates used in various types of reproduction machines; and blue, nonreproducing pencils containing a special blue lead for making nontransferable letter guide lines on drawings and for marking reports that are to be copied by certain types of reproduction machines (Fig. 1-5). The discussion that follows will address only drawing pencils and drawing leads used in drafting and will exclude the general-purpose writing pencil.

Drawing pencils need to be sharpened; that is, the wood case needs to be cut away and the lead pointed. This may be performed in one operation with a quality pencil sharpener or in two steps, as follows: The wood is cut away

Figure 1-5. Common art pencils used in various aspects of drafting for sketching and developing renderings. (Courtesy of Koh-I-Noor Rapidograph, Inc.)

Figure 1-6. Triple-duty 3-hole drawing pencil wood cutter and lead pointer. (Courtesy of Koh-I-Noor Rapidograph, Inc.)

(trimrned back) with a mat knife, pocket knife, or a single-edge razor blade, and the lead is pointed on a sandpaper block or with a hand-held lead pointing tool to provide a uniform, conical point.

The two step operation using a knife and sandpaper block is not much used in professional offices anymore. Some drafters use a hand-held combination wood cutter/lead pointer for drawing pencils while others prefer the ordinary manual or electric pencil sharpener. The combination wood cutter/lead pointer is available in either two-hole or three-hole types (Fig. 1-6). One hole is used to cut away the required amount of wood and the other for pointing the lead. This tool is easier to use, is much faster, and provides a more professional result.

Drafting lead must be kept sharp with a tapered, conical point for fine line work. This type of point can be made on a sandpaper block by rotating the pencil or lead holder while rubbing along the sandpaper surface. Hand-held, portable lead pointers are designed to provide the required shape automatically (Fig. 1-7).

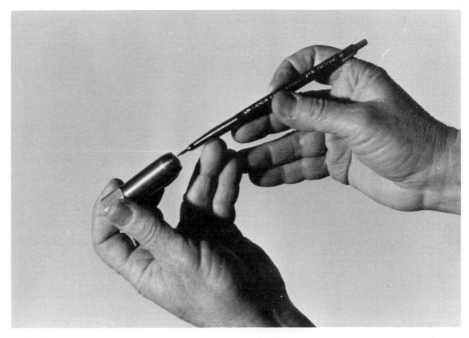

Figure 1-7. Hand-held lead pointer for use with lead holder. Hexagon shaped end lead dust collector is unscrewed to discard accumulated lead dust.

Figure 1-8. Lead pointer fastens to drafting table for one-hand operation. (Courtesy of Koh-I-Noor Rapido-graph, Inc.)

Lead pointing generates lead dust; therefore, never point lead in such a man-ner or in such a place that this free dust could settle on the drawing surface. Special care must be exercised when using the sandpaper block. Also, every time a lead is pointed, some loose graphite or polymer adheres to the point. This loose material must be removed before starting to use the pencil. Some drafters use a piece of cloth attached to the side of the drafting table to wipe the point. Styrofoam point cleaners and lubricators may be purchased for this purpose. These come in the popular ring type that fit portable or fixed lead point-ers and as small blocks with double-backed adhesive for attaching to the draft-ing table, *away from the drawing surface* (Fig. 1-8).

b. Lead Holders and Lead

In addition to the lead found in lead pencils, there are leads designed for use with lead holders and mechanical pencils. A lead holder is a mechanical device for holding specially designed sticks of drawing lead (Fig. 1-9). Lead holders are usually metal (some may have plastic bodies) that accept and hold the lead in a spring-loaded 3- or 4-prong clutch. The gripping action of the clutch is suffi-cient to prevent the lead from slipping or turning even when the lead is pointed in a blade-type pointer. A simple press of a push button at the top of the holder releases the lead for extension to compensate for wear and also for replace-ment. This device eliminates the need for cutting the wood of stick pencils.

Refill drawing leads are prepointed, approximately 5 inches long, .079/.080

Figure 1-9. Lead holder has nonslip, nonturn replaceable metal clutch that holds lead firmly in position. Button at top, when depressed, releases clutch for extending lead. (Courtesy of Koh-I-Noor Rapidograph, Inc.)

inch in diameter, imprinted with hardness degree markings for quick identification, and available in 19 degrees from 7B to 10H. One manufacturer claims that their lead diameter varies according to degree of hardness as follows:

DEGREES OF HARDNESS	DIAMETER, IN INCHES
9H through 2B	0.079
3B, 4B, and 5B	0.099
6B	0.118

Softer leads have a larger diameter for greater breaking strength.

Flat lead in degrees from 2B to 4H is available for use with flat-lead holders. Such lead is excellent for straight-line layout work as it never needs sanding or pointing.

Plastic (polymer) leads for drawing on coated polyester films are available in the more common degrees of hardness generally used in drafting.

Leads in lead holders may be pointed on sandpaper blocks or with various hand-held and fixed lead pointers.

c. Mechanical Pencils and Lead

Mechanical pencils are precision instruments for holding fine-line graphite and plastic-based leads 0.020 inches (0.5 mm) in diameter (Fig. 1-10). They are great time savers for doing fine-line drafting and lettering. The lead never needs pointing.

The leads are inserted at the cap end and automatically moved forward (extended) simply by clicking the push-button mechanism, making it easier for the user to advance the lead during work.

Mechanical pencils with features similar to the above except that they accommodate 0.036 inch (0.91 mm) (or other) diameter leads are available for heavier line work and lettering.

A fixed metal sleeve protects the lead. Some mechanical pencils have a sliding sleeve that retracts into the pencil tip gradually as the lead wears down to provide added protection for the lead during use.

Figure 1-10. Mechanical pencil for very fine drawing leads. Continuous push-button feed with positive action lead advance. Lead never needs to be pointed. (Courtesy of Koh-I-Noor Rapidograph, Inc.)

2. ERASING SHIELD AND ERASERS

An erasing shield is an indispensable device for drafters. It is a 2⅜ by 3¾-inch thin, flat piece of corrosion-resistant metal with accurately positioned cutouts. Its primary function is to permit erasure of specific line or line segments that must be deleted from the drawing. It is easy to use, and with a little practice one can become quite proficient in doing so. Since the erasing shield is flat, its handling is greatly improved by turning up one corner (Fig. 1-11).

Erasers are available in many different types with varying degrees of hardness and abrasiveness, but only a few are used in drafting. Using the incorrect eraser can damage the drawing surface. The eraser must be hard enough to remove all traces of work, without smudging, and at the same time soft enough not to destroy the drafting paper or film matte surface.

One of the more common erasers is the soft, abrasion-resistant, white vinyl eraser used for erasing lead from tracing paper and ink from drafting film without marring the drawing surface. A special characteristic of one type is the long residual strips given off during erasure that literally soak up the graphite, thus making the drawing easier to clean.

The vinyl-composition eraser is also available in a peel-off, paper-wrapped pencil type. The point of the eraser is renewed by pulling a small string downward on the peel-off case and removing the paper wrapping to expose more of the vinyl core. It is ideal for fine work and for use with the erasing shield (Fig. 1-12).

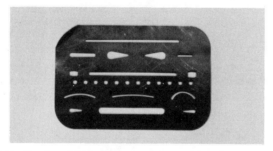

Figure 1-11. Erasing shield with one corner turned up for easier handling.

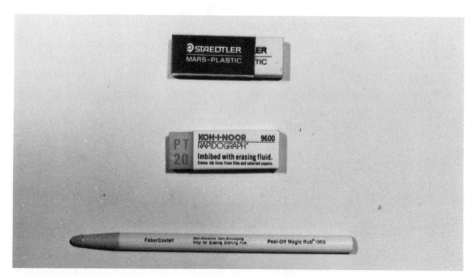

Figure 1-12. Common types of erasers used in drafting. Top–white vinyl with protective outer sleeve; center—imbibed with erasing fluid has protective outer sleeve; bottom—paper-wrapped pencil type with peel-off cover.

Kneadable rubber erasers that are molded into shape by the fingers are used to remove pencil and charcoal marks from drawings and sketches and for cleaning smudges from drawing surfaces. They may also be used as a lead point cleaner to remove loose graphite.

Electric erasers with replaceable strips are very useful when a large amount of erasure is involved. Electric erasers save time but care must be exercised so as not to damage the drawing surface.

Imbibed erasers (imbibed with erasing fluid) are especially well suited for erasing ink from polyester films with emulsion surfaces and also for erasing ink from tracing paper and vellum.

3. DRY CLEANING POWDER AND DRAFTING BRUSH

To protect drawings, a thin layer of particles is spread over the drawing surface before starting pencil work. This will protect the surface from soiling or graphite smears when scales, triangles, or T-squares are moved across the surface. This practice may not be recommended for use with some drafting films; check instructions.

Dry cleaning powder may be obtained either loose in shaker cans or in handy dry-cleaning pads. The powder is used to remove smudges from drawing surfaces and to protect such surfaces. To clean drawing surfaces, the pad is kneaded until a film of particles covers the area to be cleaned, and this area is then rubbed with the dry-cleaning pad. The same may be done with loose powder; the area is first sprinkled and then cleaned.

Drafting dusting brushes usually have a single row of soft horsehair or palma bristles with plenty of snap to remove dust particles and erasures from drawing surfaces without smudging. Removing these particles by hand or with rags tends to smear and smudge the drawings. The drafting brush is a very useful tool and should be used frequently to keep the area of work free from eraser crumbs and graphite dust. Drafting brushes must be kept clean or they will contribute to the problem. Most drafting brushes can be cleaned with soap and water and then left to dry thoroughly before using.

Figure 1-13. Drafting brush should be used to remove lead particles from drawing and to keep drafting table clean. Dry-cleaning powder in convenient handy sack.

4. PEN AND INK

There are several different types of devices used for ink work in drafting. These include ruling pens (considered to be a part of drawing instrument sets), technical pens, and interchangeable-nib drawing fountain pens. Each has its place in drafting, and some offer more versatility than others.

The next step after the student of drafting has developed sufficient proficiency with lead on paper is to learn to work with ink. There are several distinct advantages of ink drawings over lead. For example, pens do not need to be sharpened or pointed (ruling pens may need honing after extended use); lines of different widths can be drawn with one stroke whereas several strokes may be required with lead; ink will not smudge once it has dried; reproduction of ink drawings usually have more uniform line quality, and ink on polyester film produces permanent drawings.

a. Ink

Ink used for drafting is formulated either for general-purpose use or for specific instruments and drawing mediums. Microfine carbon particles are used as the pigment, and the vehicle varies according to the intended use.

General-purpose inks are those formulated for use with cartridge technical pens, direct filling ruling pens, and reservoir lettering pens for polyester film, vellum, paper, and cloth. These inks also work in automated drafting machines, are free-flowing, fast drying, have permanent adhesion, and yet are easily erasable from drafting film.

Some manufacturers produce special-purpose ink formulated for use on polyester drafting film. Waterproof formula clings to drafting film and won't peel, chip, or crack. Opaque matte finish of this ink won't reflect light and reproduces without hot spots or line variation.

Inks are also formulated for exclusive use in automated drafting machines for plotting on film, paper, or cloth. These have long open-pen time to keep the ink free-flowing at all times.

One product has a latex binder which permits a high carbon concentration for optimum blackness and opacity while maintaining the free-flowing characteristic of ink.

Acetate ink is for use on clear acetate, coated and uncoated film. It is especially useful for preparing overhead projector transparencies since it easily wipes clean with alcohol or water from acetate or plastic transparencies and thus facilitates redraws to illustrate progressive changes.

A good grade of universal drawing ink will satisfy most normal drafting needs. Drawing inks are usually packaged in plastic bottles with a filler spout or dropper stopper for convenience in filling pen cartridges and for direct filling of ruling pens.

b. Drawing Instruments

Drawing instruments are essential tools for drafters. They are available individually or in complete sets. The sets come in student economy quality or sturdy, precision instruments engineered for rugged durability, extreme accuracy, and efficiency in performance for the experienced professional (Fig. 1-14).

Drawing sets consist of ruling pens, compasses and dividers, with a variety of attachments and adaptors for versatility. For example, compasses may be specifically designed to hold lead or may be designed for more universal use to

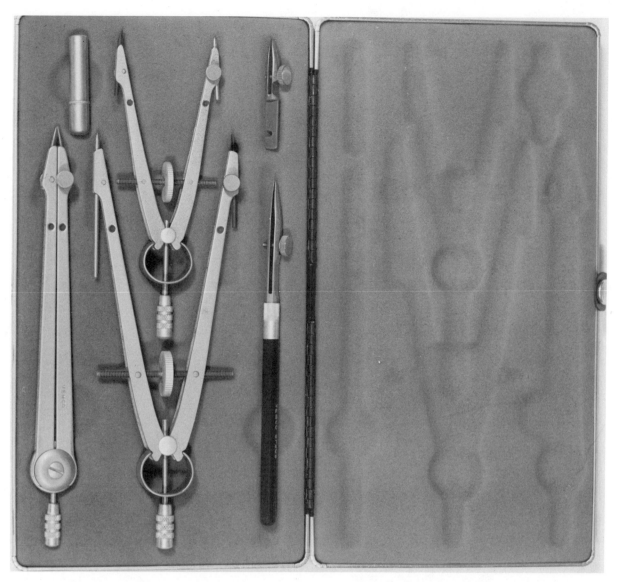

Figure 1-14. Drawing instrument set consists of: ruling pen and pen attachment; large and small bow compasses; friction divider; and a container of extra compass leads and divider points. (Courtesy of VEMCO Corporation)

accept attachments for use with lead and ink. There are also beam attachments for compasses to extend their use for larger circles. Compasses also have quick-set features for rough setting plus microscrew for fine adjustment (Fig. 1-15).

Dividers are similar to compasses, except that they have friction adjustment instead of screw and have metal points on both legs. They are used to lay off distances and to transfer measurements. Proportional dividers are double-ended with four points, have sliding, adjustable pivots that vary the proportion of their legs, and are used to divide lines and circles and transfer reduced or extended measurements.

The ruling pen is a special device generally used for line work. It consists of a pair of adjustable, parallel metal prongs attached to a plastic or metal handle. The gap between the prongs is adjusted with a thumb screw, and the gap determines the width of the line that will be drawn. The ruling pen is a universal tool that, with a simple turn of the thumb screw, the prongs can be adjusted from close together for a fine line to wide apart for a thick line.

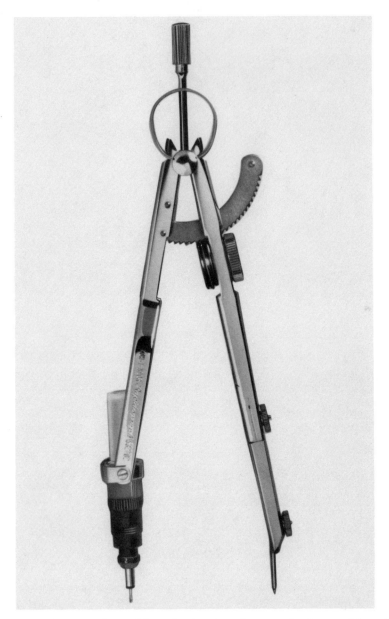

Figure 1-15. Ringhead compass has quick-set feature plus microscrew for fine precision adjustment with attachment for technical pen. (Courtesy of Koh-I-Noor Rapidograph, Inc.)

The pen is filled by placing a few drops of drawing ink from bottles with filler spout or dropper stopper into the space between the prongs. Any desired line thickness can be drawn by merely adjusting the distance between the prongs.

To become proficient in the use of the ruling pen requires plenty of practice. The pen is held perpendicular to the drawing surface and tilted slightly in the direction of the draw. A gentle pressure is all that is necessary to draw a clean, uniform line. Do not permit the pen to remain motionless at the start or end of a line, as excessive ink will flow. Start freely and end cleanly.

A few words of caution about the ruling pen. Do not over tighten as the prongs may become deformed. Always clean the ink from in between and on the prongs when the pen is not in use. Leave the prongs slightly open when the pen is stored.

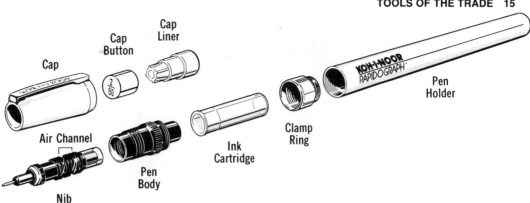

Figure 1-16. Disassembled view of a technical pen shows replacement parts and miscellaneous items. (Courtesy of Koh-I-Noor Rapidograph, Inc.)

6x0	4x0	3x0	00	0	1	2	2½	3	3½	4	6	7
.13	.18	.25	.30	.35	.50	.60	.70	.80	1.00	1.20	1.40	2.00
.005 in.	.007 in.	.010 in.	.012 in.	.014 in.	.020 in.	.024 in.	.028 in.	.031 in.	.039 in.	.047 in.	.055 in.	.079 in.
.13 mm	.18 mm	.25 mm	.30 mm	.35 mm	.50 mm	.60 mm	.70 mm	.80 mm	1.00 mm	1.20 mm	1.40 mm	2.00 mm

Figure 1-17. Technical pen point chart shows relationship between pen point number and line width from #6 × 0 to #7. (.005 inch-.13 mm to .079 inch-2.00 mm) (Courtesy of Koh-I-Noor Rapidograph, Inc.)

Technical Pens

The technical pen is a versatile drawing instrument for use with various inks, handles like a pencil, and can be moved in any direction without snagging or digging into the drawing surface. It is available with up to 13 different, color coded point sizes ranging from 0.13 mm (No. 6 x 0) to 2.00 mm (No. 7) and three point materials for various drawing surfaces and personal preferences. With technical pens, the line width depends upon point size, one line width for each point size and the drawing cones are identified by a number and are also color coded.

The metric point size designations are in millimeters. Nine line widths conform to International Standards Organization and are compatable with the standard American drawing sizes. The line widths provide geometric progression in the square root of 2 (1.414) to facilitate "half-scale blow back." These are 0.13, 0.18, 0.25, 0.35, 0.50, 0.70, 1.00, 1.40 and 2.00 mm (Fig. 1-17).

Points are currently available in three materials:

1. Stainless steel recommended for use on cloth, tracing paper and vellum.

2. Tungsten carbide developed primarily for use with programmed automated drafting machines to provide optimum abrasion resistance on coated drafting film and for hand drafting on film.

3. Sapphire jewel is self-polishing and remains smooth on the abrasive surface of coated drafting film and may also be used on paper or cloth. It has been life tested to last up to 500 times longer than stainless steel points.

Technical pens consist of removable drawing cones (drawing cone contains the point), see-through refillable ink cartridge, pen holder and specially designed, airtight cap that allows instant start up for increased productivity.

To fill, merely unscrew holder, remove ink cartridge and fill with appropriate drawing ink to no more than ¾ full. A few gentle "shakes" of the assembled pen usually starts ink flow.

Figure 1-18. Technical pen is held vertically for drawing lines and lettering. (Courtesy of Koh-I-Noor Rapidograph, Inc.)

The pen is held perpendicular to the drawing surface and perpendicular in the direction of draw. It is used for line work, dimensioning and lettering. The cap must be replaced when the pen is not in use to prevent ink from drying in the point. In the event the point is damaged, replacement drawing cones may

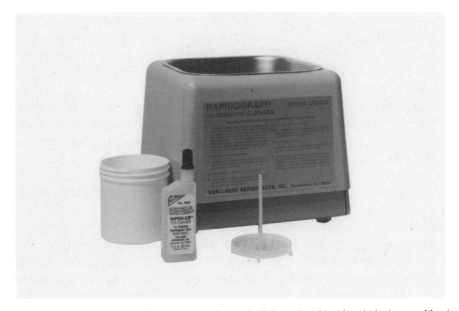

Figure 1-19. Ultrasonic cleaner provides nonabrasive and efficient cleaning of technical pens. May be used for points alone or fully assembled pens. (Courtesy of Koh-I-Noor Rapidograph, Inc.)

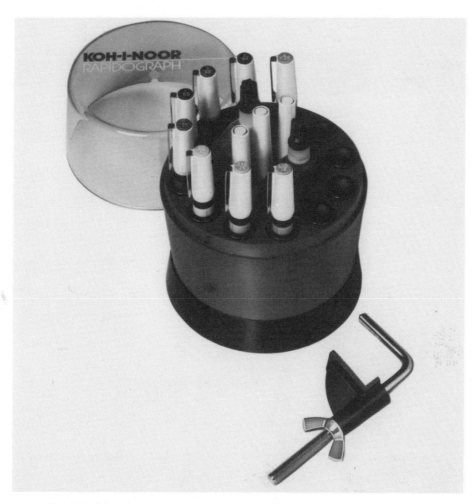

Figure 1-20. Humidified interior to technical pen humidifier prevents ink from drying when points are stored uncapped, point down, in receptacle. (Courtesy of Koh-I-Noor Rapidograph, Inc.)

be purchased individually. Pens should be stored in the vertical position with the point end up and the cap in place.

Ultrasonic cleaner may be used to clean fully assembled technical pens and individual drawing cones. A humidifier prevents ink from drying in the point. During normal working periods when using more than one pen, idle pens may be stored in a humidifier, uncapped, point down. The points are exposed to humidified atmosphere within the unit, but not to the wetting agent (Fig. 1-20).

Liquid pen cleaner may be used to unclog technical pen points. In severe cases when flow cannot be started by shaking the pen, the drawing cone is removed and placed in the liquid solution overnight or as long as necessary to free the point. Pressure syringes may be used for cleaning points and to start ink-flow.

d. Interchangable Nib Fountain Pen

This is a drawing ink fountain pen with easily interchangeable stainless steel nibs. The drawing ink flows uniformly and continuously from the reservoir through the nib when in use, and gives a clean, sharply defined line (Fig. 1-21).

Nibs are available in a wide range of styles (pointed, rounded, and tubular) and sizes for line work, sketching, lettering, and free-hand drawing. Nibs can be quickly and easily mounted and removed (Fig. 1-22, Fig. 1-23).

Figure 1-21. Interchangeable nib technical fountain pen. (Courtesy of Koh-I-Noor Rapidograph, Inc.)

Figure 1-22. Eight different types of interchangable nibs for a variety of line work, sketching, and lettering. (Courtesy of Koh-I-Noor Rapidograph, Inc.)

Follow instructions for filling the pen reservoir. Filling may be done by removing the "feed piece". Exhaust some of the ink from the reservoir with an empty pipette, when overfilled, since otherwise ink may flow out when the feed piece is inserted. The reservoir may be filled without removing the feed piece by dropping ink into the opening in the feed piece. Fill to the level of the fill hole.

For line work the pen is held perpendicular to the drawing surface and at approximately 60° in the direction of the draw. For lettering, use as any ordinary fountain pen. The pen may also be used with a drawing instrument set compass with a compass clip adapter.

5. TRIANGLES, TEMPLATES, SCALES AND OTHER DRAWING AIDS

Drawing aids are useful and/or necessary tools specifically designed to simplify and expedite drafting. Triangles, templates, french curves, protractors, and lettering guides are usually made of rigid, transparent clear or colored (green or orange) plastic in thickness of from 0.03 to 0.10 inch, depending upon size and use.

There are two types of triangles, fixed and adjustable. Fixed triangles are available in the 60°/30°/90° and 45°/45°/90°, ranging in size from 4 inches to

Nib Assortment Chart

Pelikan Graphos Interchangeable Nibs for use with Graphos technical fountain pens are available in 60 different styles. The nibs, with the exception of the flexible "S" type, are of a predetermined width not affected by varying pressure. Graphos Nibs are rust-resistant. Their pivot construction permits easy and thorough cleaning.

ITEM NUMBER	NIB	TYPE	WIDTHS SUPPLIED (in mm)	KIND
9010-A		A	0,1 0,13 0,18 0,2 0,25 0,3 0,35 0,4 0,5 0,6 0,7	RULING NIBS FOR FINE LINES
9010-T		T	0,8 1,0 1,25 1,6 2,5 4,0 6,4 10,0	RULING NIBS for broad lines and for poster work
9010-S		S	B = SOFT HB = MEDIUM HARD H = HARD K = EXTRA HARD	Drawing Nibs for fine freehand drawing sketching, cartography and touch-up work
9010-R		R	0,3 0,4 0,5 0,6 0,7 0,8 1,0 1,25 1,5 1,75 2,0 2,5 3,0	TUBULAR NIBS FOR STENCILLING WITH LETTERING GUIDES AND FOR CONTOUR LINES
9010-O		O	0,2 0,3 0,4 0,5 0,6 0,7 0,8 1,0 1,25 1,6 2,0 2,5 3,2 5,0	ROUND NIBS for freehand lettering and sketching
9010-N		N	0,8 1,25 2,0 2,5 3,2 4,0	Right hand slant nibs for oblique lines
9010-Z		Z	0,8 1,25 2,0 3,2	Left hand slant nibs for oblique lines

		Unit Pack			Unit Pack
9010-A	Ruling nibs for fine lines.	12	9010-N	Right-hand slant nibs for squared end lines.	12
9010-T	Ruling nibs for broad lines & poster work.	12	9010-Z	Left-hand slant nibs for squared end lines.	12
9010-S	Drawing nibs for free-hand drawing.	12			
9010-R	Tubular nibs for lettering guides.	12	9006-01	Cleaning wires for tubular nibs "R" for 0.3 mm.	25
9010-O	Round nibs for rounded end lines.	12	9006-02	Cleaning wires for tubular nibs "R" for 0.4 to 3.0 mm	25

Figure 1-23. Nib assortment chart. 60 interchangeable stainless stell nibs for ruling, freehand lettering, calligraphy, cartooning, and poster work. (Courtesy of Koh-I-Noor Rapidograph, Inc.)

A 60° triangle is measured on the longest cathetus, LC, and a 45°
is measured on the length of the two equal sides, C.

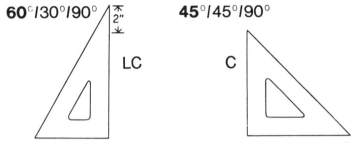

Two triangles are considered a set if LC for a 60° is two inches longer than C
for a 45°.

Figure 1-24. A set of fixed triangles. (Courtesy of Koh-I-Noor Rapidograph, Inc.)

about 20 inches as measured on the longest cathetus, LC. A 60° triangle is
measured on the LC, and a 45° is measured on the two equal sides, C. Two
triangles are considered a set if the LC of a 60° is 2 inches longer than C for a
45° triangle. For the student a set with an LC of 12 inches is usually sufficient.
Professionals frequently have several sets, one small, one medium, and one
large (Fig. 1-24).

The adjustable triangle has a closed position of 45° and can be opened and
locked into any position up to 90° (Fig. 1-25).

A set of fixed triangles used alone or in combination can provide 15°, 30°,
45°, 75°, 90°, 120°, 135°, 150°, 165°, and 180° (any angle divisible by 15°).
The adjustable triangle can be set at any angle from 45° to 135° when used
alone and from 0° to 180° when used with another triangle.

The student should have one set of fixed and one adjustable triangle. It is

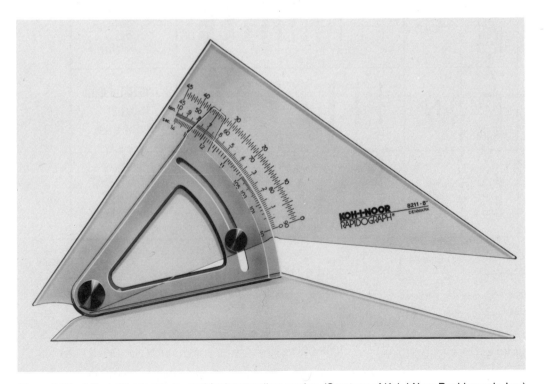

Figure 1-25. Adjustable triangle is used for intermediate angles. (Courtesy of Koh-I-Noor Rapidograph, Inc.)

recommended that fixed triangles be used for all angles that can be obtained from it and that the adjustable triangles be used for intermediate angles.

Templates are drafting aids designed to speed the work of drafters and to ensure accurate, repetitive designs of all normal shapes and almost any conceivable symbol. Standard shapes include the circle, triangle, hexagon, and the square. Templates are available with various sizes of one shape on one template, i. e., all circles, or with all four shapes on one template. Other shapes include the ellipse and the isometric ellipse. Symbols run the gamut from landscape, architectural, office furniture and equipment, plumbing fixtures, pipe fittings, flow charts, electrical equipment, electronic controls for computer logic, and math symbols (Fig. 1-26, Fig. 1-27).

A starter set should include one combination circle/hexagon/square/triangle, one with all circles, and symbol templates depending upon the field of drafting pursued. For ink work, edge strips are available that elevate the template to eliminate ink smears.

Reduction scales, commonly known as scales, are made in several sizes, shapes and materials. The 12-inch scale is generally used for board work and the 6-inch, with a leather case, is conveniently carried in the pocket for scaling drawings in the field, and may be used for making detail drawings where the 6-inch size is adequate. Scales larger than 12 inches are available. The scales are made of several shapes: triangular, four-bevel, and two bevel flat scales, and may be solid wood, solid plastic, and basswood with plastic edges.

Two types of scales are used in building construction drafting and design: architect's scale, and civil engineer's scale, commonly referred to as the engineer's scale, and is sometimes called a *decimal inch scale* (Fig. 1-28).

Architect's scales are used for making building drawings and are *open-divided* scales, that is, only one unit of the scale is fully divided and the remaining units are open or free from subdivision. The open section is used to measure full foot dimensions and the subdivided section is used to measure inches and fractions of an inch.

These scales are easy to use, for example, on the scale where $\frac{1}{4}$-inch equals one foot, each major mark to the left of zero represents one foot. To measure 20 feet, merely read left from zero to the line marked 20. The face of the scale $\frac{1}{4}$-inch per foot to the left, is also marked to read $\frac{1}{8}$-inch per foot to the right. Do not confuse the numbering system for the two different scales. The fully divided section on the $\frac{1}{4}$-inch per foot face to the right of zero has 12 subdivisions with each mark representing one inch. To read 20'-6", read from the 20 mark to the 6th mark past zero. Other scale faces have a different number of subdivisions outside the zero mark, depending upon the scale involved and the space for legible markings. For example, on the face where 1-inch equals one foot, there are 48 subdivisions, each representing $\frac{1}{4}$-inch. Each face of an architect's scale has two companion scales except one, and that is either a 6-inch or 12-inch rule, divided in $\frac{1}{16}$ of an inch. The companion scales are: $\frac{3}{32}$-$\frac{3}{16}$, $\frac{1}{8}$-$\frac{1}{4}$, $\frac{3}{8}$-$\frac{3}{4}$, $\frac{1}{2}$-1, $1\frac{1}{2}$-3. Each face is identified outside the zero mark as to its scale.

Engineer's scales are generally used for civil engineering work, site plans, hence the name, civil engineer's scales. They are fully divided and the measurements are read directly from the scale. The subdivisions between the numbers on the face represent decimal fractions of a foot, for example, on the face marked 10, each numbered mark represents one foot and each subdivision equals $\frac{1}{10}$ of a foot. Engineer's scales have one scale per face and include $\frac{1}{10}$, $\frac{1}{20}$, $\frac{1}{30}$, $\frac{1}{40}$, $\frac{1}{50}$, and $\frac{1}{60}$.

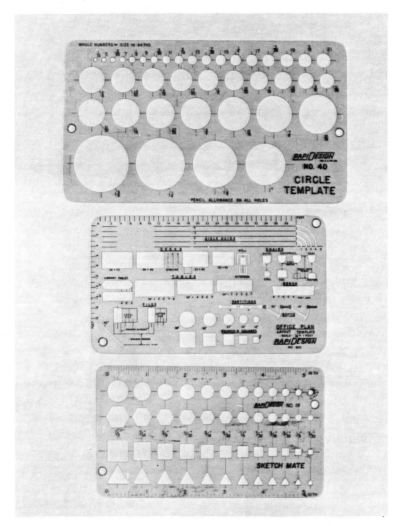

Figure 1-26. Common templates used for building construction drafting. Top—circle template; center—office plan; bottom—sketch mate.

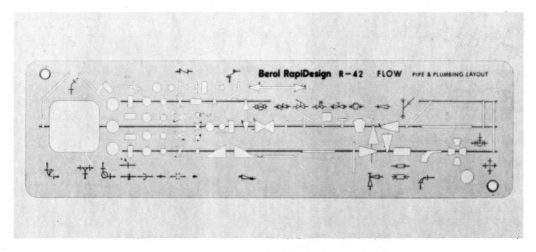

Figure 1-27. Pipe and plumbing layout template.

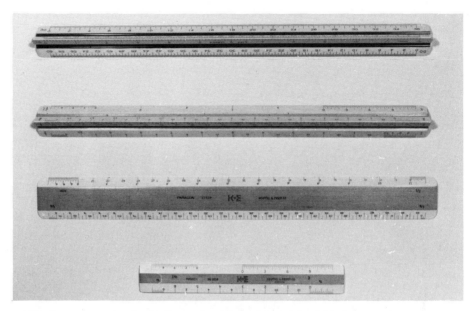

Figure 1-28. Common types of reduction scales. Top—12 inch fully divided engineering triangular scale divided into 10, 20, 30, 40, 50, and 60 parts to the inch; next—12 inch open divided architectural triangular scale divided into ³⁄₃₂, ³⁄₁₆, ⅛, ¼, ½, 1, ⅜, ¾, 1½, 3 inches to the foot, one edge divided into 16ths; next—12 inch two-bevel flat scale divided into ⅛, ¼, ⅜, ¾, ½, 1, 1½, 3 inches to the foot; bottom—6 inch four-bevel pocket scale divided into ⅛, ¼, ⅜, ¾, ½, 1, 1½, 3 inches to the foot.

French curves are made of material similar to triangles and templates and are available in a variety of curved surfaces. They are used to produce compound curved lines that cannot be made with templates or compasses (Fig. 1-29).

Since many compound curves require more than one setting of a french curve, or more than one french curve, the construction of a uniform line that will appear as if made in one stroke is rather difficult. The proficient use of french curves require diligent practice.

Protractors are also made of transparent plastic and are available in semi-circular and circular types. The semi-circular protractor is marked off into 180° reading from zero to 180° in both directions. The circular type is marked off into 360° reading from zero at the horizontal base line and from zero to 360° in both directions. This dual reading permits angles to be measured and layed out in any direction quickly and easily (Fig. 1-30).

While the protractor is a useful tool for accurate work, the adjustable triangle is frequently used for the same purpose.

There are several different types of lettering guides on the market, but they all perform the same function. Lettering guides may be "fixed" (triangular) or adjustable (circular). Both are provided with several sets of holes into which the pencil point is inserted and guide lines are made by sliding the guide along a T-square or parallel straightedge. The guides also have a slot used to draw inclined letter guidelines (Fig. 1-31, Fig 1-32).

Lettering guidelines are necessary for the construction of uniform sized lettering and numbering. Even experienced professionals use them. Nothing detracts more from a well designed drawing than poorly spaced, nonuniform sized and sloppy lettering.

Lettering guides usually have complete instructions included that explain how to use them. Buy one and learn to use it proficiently.

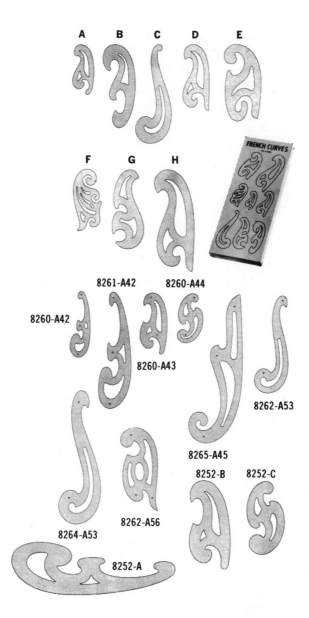

Figure 1-29. Assorted clear plastic french curves available individually or in sets.

Guidelines should be light lines drawn with a well pointed hard pencil so as not to interfere with clarity of the lettering. A blue lead may be used to draw guidelines, it serves as a legible guide but will not reproduce when the drawing is copied.

6. DRAFTING EQUIPMENT

Drafting equipment includes T-squares, parallel straightedges, drawing boards, drafting tables, and drafting machines. There are several manufacturers of these products, and most make various grades that can be used at home, in school, or in professional offices.

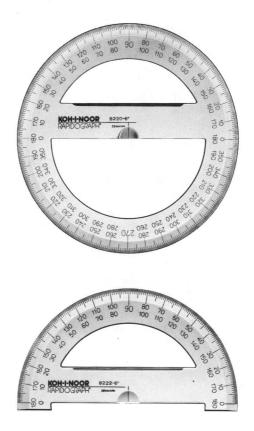

Figure 1-30. Clear plastic circular and semicircular protractor. (Courtesy of Koh-I-Noor Rapidograph, Inc.)

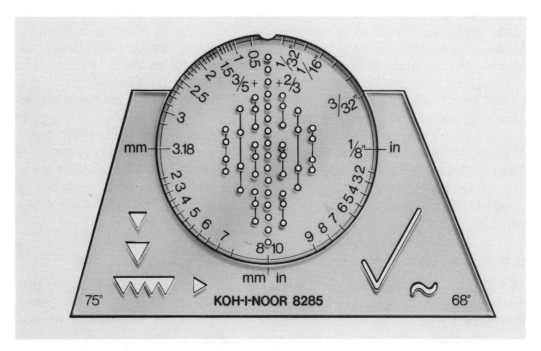

Figure 1-31. Adjustable lettering guide has a transparent plastic disc in a plastic frame. (Courtesy of Koh-I-Noor Rapidograph, Inc.)

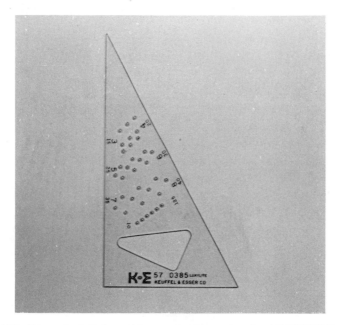

Figure 1-32. Triangular lettering guide may also be used as fixed 60°/30°/90° triangle.

a. T-square

The T-square is a portable device used on drawing boards and drafting tables for constructing parallel lines and serves as a base for triangles, etc., when lines other than horizontal ones are required. It consists of a head and a perpendicular blade. The blade may be of wood (maple, beach) with clear transparent edges that should be raised or undercut so that it will not touch the drawing surface. This slight elevation discourages capilliary action when inking. The blades are fastened to the head with screws and are generally available with blade length sizes from 18 to 48 inches. Some T-squares have black phenolic blades with transparent edges and a demountable hardwood head. There are also metal T-squares but these are not usually used for drafting.

T-squares must be handled with care so as not to disturb alignment, and should be checked to ensure parallel line work. Screws should also be checked at regular intervals to prevent loosening.

In use, the T-square head is held firmly against the side of the drawing board or table and is slid along that edge to construct parallel lines. The transparent edge permits the drafter to see some of the drawing below the area of work. The transparent edges are undercut, or raised so that they will not touch the paper as it is moved up and down the board to eliminate "smears" when doing ink work.

b. Parallel Ruling Straightedge

The parallel ruling straightedge is similar in function to a T-square, except that it is permanently fastened to the drawing board or drafting table. It has undercut transparent plastic edges and the blade is generally made of dimensionally stable black phenolic laminate. In cross section, it is thicker than the T-square to accommodate cables and four ball bearing pulleys.

The straightedge is attached to the board or table by a cable through a system of pulleys to maintain its parallel position while it is moved up and down. The

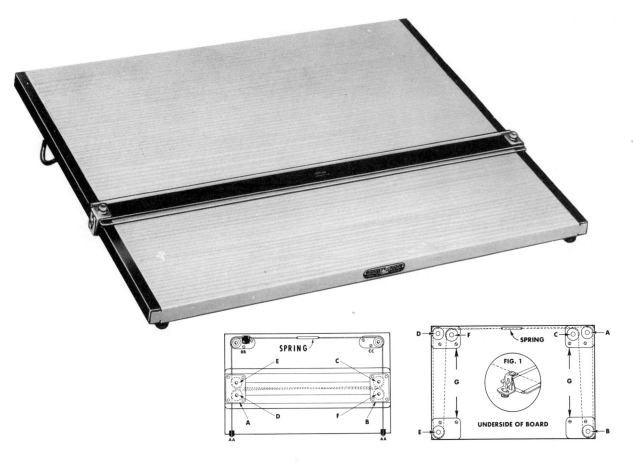

Figure 1-33. Mayline professional drawing kit includes solid basswood, metal edge drawing board with clear lacquer finish and parallel ruling device. Phantom views show cable and blade attachment for parallel ruling straightedges. Left—above board attachment; right—underboard attachment, cable on underside of board. (Courtesy of MAYLINE Company, Inc.)

cable may be run over or under the board. Some straightedges have a breaking mechanism for locking it into position for use with lettering sets or inking. A counterbalance may be added when necessary with tables used at highly elevated angles.

c. Drawing Boards

Drawing boards are portable devices made of kiln-dried basswood, about ¾ inch thick, with or without steel end cleats. Because lumber is a living substance, drawing boards should be stored in a cool dry place to ensure complete stability. They are available in sizes from about 12 by 17 inches to 30 by 42 inches and may be used with a T-square or parallel ruling straightedge that have cable attachments above or under the board (Fig. 1-33).

d. Drafting Tables

The pedestal drawing table is made from kiln-dried hardwood and is ideal for home, school, factory office, field office, trailer, art studio, or wherever a table is needed. It is available with or without a parallel ruling straightedge. Table tops

Figure 1-34. Sturdy, professional quality pedestal drafting table made from kiln-dried hardwood. Basswood drafting top, height adjustable from 30 to 40 inches. (Courtesy of MAYLINE Company, Inc.)

are generally 24 by 36 inches wide, adjustable from 30 to 40 inches in height, and may be tilted from parallel to the floor to approximately 80° from horizontal. (Fig. 1-34).

Drafting tables are made with wood or metal bases, with or without drawers, with manually adjustable top slope or with fully automatic, spring loaded with fingertip control for automatic tilt angle and top height adjustment. These tables are usually used in professional offices. They may be purchased with or without parallel ruling straightedge or drafting machine (Fig. 1-35).

While drawing board surfaces are generally not "covered", drafting tables frequently have drafting linoleum or vinyl covers to protect the surface and to provide a smooth resilient surface for drafting. These coverings are attached with double-faced tape along the top edge of the table. NEVER cement this material to drafting tops, since this can cause the wood to warp and the surface material may become irregular, and in the event of damage the covering cannot be easily replaced.

Dust covers should be used on drafting tables to protect drawings during idle periods and from prying eyes. Drafting tables usually come with a pencil trough along the bottom edge of the board to provide a convenient receptacle for pencils, pens, scales, erasers, etc.

Another accessory is the drawing protector. It is a circular metal tube about 2 inches in diameter and is fastened along the bottom edge of the table. By sliding the drawing down into the drawing protector, the bottom section is coiled safely out of the way and keeps the drawing free from elbow smudges, wrinkles, and torn edges while working for extended periods near the top of the drawing.

Figure 1-35. Steel 4-post drafting table with plan and tool drawer. Basswood top with steel end cleats, height 37 inches from floor. (Courtesy of MAYLINE Company, Inc.)

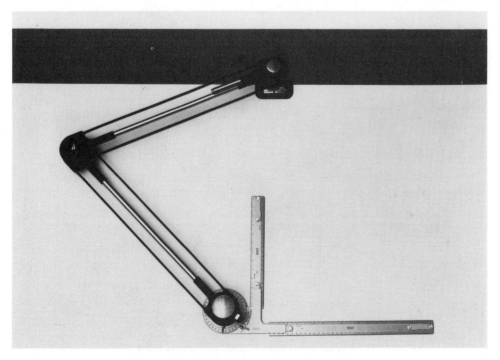

Figure 1-36. VEMCO elbow-type drafting machine combines the working features of a T-square, a protractor, various scales, and triangles in a single unit. (Courtesy of VEMCO Corporation.)

e. Drafting Machines

A drafting machine is a device that attaches to the drafting table and combines the working features of a T-square (or parallel ruling straightedge), a protractor, various scales and triangles into a single unit, ready for immediate use, while leaving the drawing surface clean for working. They are built for speed and accuracy making architectural and engineering drawings, and for general all around drafting.

There are two types of drafting machines in common use today; the arm type and the track type.

The arm (elbow) type, band and pulley machines are recommended for smaller drawings on flat or moderately inclined boards. They are available in several sizes with board coverage of 30 by 60 inches, 42 by 84 inches and 48 by 96 inches (Fig. 1-36).

The protractor (control) head contains a metal bracket to hold two drafting machine scales at right angles to one another. The scales are made of select quality, clear, high-impact plastic, slightly tinted to reduce glare and perma-

Figure 1-37. VEMCO V™-Track drafting machine for use on larger tables with tilt angles up to 50–60° from the horizontal. (Courtesy of VEMCO Corporation.)

nently molded to a hardened and bronze-anodized aluminum rail. They can be quickly and simply inserted and removed from the holder.

The drafting head has a protractor and automatic indexing device for setting scales at any desired angle with notch settings at 15° intervals and independent positions for any angle.

Drafting machine scales are offered in sturdy crystal clear plastic with precision graduations fully divided to eliminate frequent shifting. Graduations and numerals are on the lower surface of the scale preventing parallax errors.

A supporting center rib raises graduated area to prevent scratches and wear, and discourages capillary action when inking. Scales are available in several lengths with the usual graduations for architectural and engineering work. Scales are designed to be changed simply by special metal clips fixed to the scales for insert into metal brackets on the protractor head.

The arm-type of drafting machine works well on boards tilted up to 10° or with gravity-compensating, adjustable counterpoise device up to 20°.

The track type of drafting machine has both horizontal and vertical track to guide the protractor head to any required position on the board. These are designed for larger boards up to 37 by 72 inches. Chalk board drafting machines are available for chalk boards up to 48 inch by 16 feet. They are also well suited for tilted boards up to 50–60° from the horizontal (Fig. 1-37).

The protractor head has all the normal functional features described under the arm-type.

Drafting machines are precision instruments and should be treated accordingly. They should be checked for accuracy at regular intervals in accordance with manufacturer recommendations.

Chapter 2
Basic Drafting Skills

Basic drafting skills are essential for the development of quality building construction drawings. These drawings must be clean, the line work sharp, and the lettering legible. Drafters must learn to use mechanical aids where necessary for hard line work and should become proficient in freehand sketching. Once basic drafting skills are developed, drafters need to learn how to use those skills for making drawings. Anyone can learn to make straight lines with a straight-edge and curved lines with a french curve, but this is only the beginning. The next step is to be able to combine these various lines to properly and adequately illustrate a building and all its essential components. This too can be learned, with practice. While quality line work is essential, neat and legible lettering is far more important. Lettering is the element that gives meaning to the myriad of lines that graphically depicts the building structure. Well executed lettering contributes more than any other single factor to the development of quality building construction drawings, conversely, sloppy lettering detracts from a good hard line drawing.

Architectural and engineering drawings are usually developed in three phases: schematics, preliminaries, and working drawings. While all three phases of drawing development are important, the first two are usually development drawings, and unfortunately, all too frequently, less attention is paid to exacting details. The working drawings, also known as construction drawings, and are, as the name implies, *construction drawings.* These are used by the contractor to erect the structure, and for that purpose, the line work must be sharp and the lettering legible.,

Anyone interested in pursuing a career in building drafting and design must develop basic drafting skills. This includes freehand and hard line work, and freehand and mechanical lettering. In building construction drafting, sketches are usually freehand and drawings are hard line; that is, line work is made with a mechanical aid.

1. DRAFTING TECHNIQUES

Building construction drawings consist of a wide variety of straight and curved lines assembled in such a fashion that the end result is the *picture* of a building. Freehand lines may be used for sketching and schematic drawings, but preliminaries and working drawings consist of hard line work. Hard line drawings refer to drawings made with the aid of mechanical devices such as a ruling straightedge, T-square, triangles, french curves, etc.

Drafters and designers are not artists, but artistic ability is helpful and good finger dexterity is very useful. The beginning drafter should develop an under-standing of fundamental drawing techniques. Most importantly, proficiency comes with practice, and this is a two step process; first learning the basics of good line work and then developing speed. The person that can draw well, and rapidly, will always have opportunities for employment.

a. Freehand Drafting

Various types of freehand work are necessary in building construction drafting. For example, sketches are needed to develop concepts quickly, and are made during site visits for renovation and new construction projects. Architects and engineers use freehand sketches from which drafters make hard line drawings.

Freehand drawings are often done on graph paper with either square or iso-metric patterns as an aid for developing sketches. Graph paper is available in a variety of patterns. The square pattern is used for plan views and elevations, while the isometric pattern is used for piping drawings, riser diagrams, equip-ment layouts, etc., where three dimensional views are necessary. Graph paper is also a valuable aid for sketching to scale. The graph paper with eight subdi-visions per inch can be used for making ⅛″ = 1′-0″ sketches directly on a sketch pad without the need for architectural scales. Since each ⅛th inch subdivision represents one foot, drafters can quickly make sketches that are to scale.

Freehand sketches and drawings are made with a sweeping motion of the hand, the entire arm slides along the paper with fixed wrist position. The begin-ner should practice with either graph or lined paper, using short strokes at first. The practice exercise should include horizontal, vertical and slanted lines. At the outset, all the lines should be short, about one to two inches long. Once this is mastered, longer lines can be drawn. For the right handed person, all lines are drawn from left to right with the pencil slanted slightly in the direction of the draw. Mechanical pencils and technical pens are held in the vertical position for all direction of motion. Slanted lines are drawn from left to right from 0° through 90°, and as the angle increases, the lines will be from the bottom upwards. From 90° through 180° the lines are drawn from the top down until at 180° the line is again horizontal and is drawn from left to right (Fig. 2-1).

Freehand sketches are usually made with soft lead, H, F, or B. For preliminary sketch work, some architects use carbon pencils on sketch paper since this will

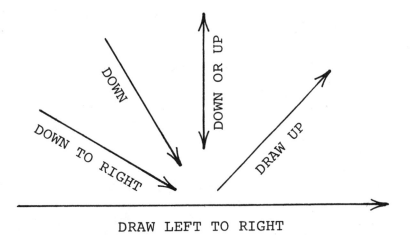

DRAW LEFT TO RIGHT

Figure 2-1. Right-handed drafters draw all lines away from the body, that is, left to right.

produce a broad, dark line with very little pressure on the stroke. For sketches that require finer lines and dimensioning, some people may even use a 2H lead. The choice of lead grade is usually a personal preference, and depends upon the quality of the drawing medium used. Light weight sketch paper may tear if the lead is too hard. Freehand sketching is a valuable aid in building construction drafting.

Freehand sketching, like freehand lettering, requires diligent practice and conscientious effort. There are certain basic guidelines the beginner drafter should follow to develop an adequate technique. Sit or stand in a comfortable position at the table with weight evenly distributed and the back fairly straight. Place both arms on the table so that the forearms touch the surface and the arms form a triangle on the table. This is more important for sketching than lettering as the strokes are longer for sketching. Short horizontal strokes are made with a wrist motion and longer strokes are made with a fixed-wrist, full-arm motion. Short vertical strokes are made using finger motion but for longer strokes the wrist is fixed and the arm slides along the table.

Hold the pen or pencil firmly but not tightly and the strokes should be a free and easy motion.

Figure 2-2 is an exercise sheet for the beginner. Use square grid graph paper and practice drawing ON the background lines. Start with short strokes in the direction of the arrows and work up to longer lines as proficiency develops. After single lines can be drawn fairly straight, try drawing parallel lines as these are often required in building construction drafting.

This exercise should be practiced at first with a soft lead pencil or a carbon sketching pencil. Beginners develop confidence faster with broad strokes than with fine lines.

Freehand lines appear different from instrument drawn lines. As long as the lines are relatively straight and parallel (for two lines), do not be overly concerned about slight squiggles. Figure 2-3 shows freehand sketch of site plan.

Another excellent technique for developing freehand drafting skills is the use of white or canary sketching paper over a drawing and tracing the building outlines or details freehand. This sketch paper is economical and comes in rolls of 12 to 42-inches wide by 50 yards long. In addition to aiding the beginner to develop freehand skills, it also provides practice for making building drawings. In addition to soft lead pencils, nylon or fiber tip pens may be used for freehand sketching and overlay drafting. The 8-pound lightweight sketch paper (for overlay) is especially receptive to soft pencils, felt tip pens, and ink. A wide variety of sketching techniques can be performed with this type of paper at a very reasonable investment in material.

Once the freehand line strokes have been mastered the drafter is ready to practice sketching buildings and details on 8½-inch gridded paper. At this point the drafter should begin to use the appropriate line type for showing the different aspects of the building. Lines of various shapes (dotted, dashed, etc.) and weights (thicknesses) have different meanings. For example, a thick line is usually used for the building outline and fine lines are used for dimension lines (see Figure 5-9).

Isometric graph paper is used for making freehand piping drawings, riser diagrams, and equipment layouts. A drawing on isometric paper may be scaled in three main directions, the axes of which are 120° apart, one being vertical and the others being 30° from the horizontal. All horizontal lines are laid along the

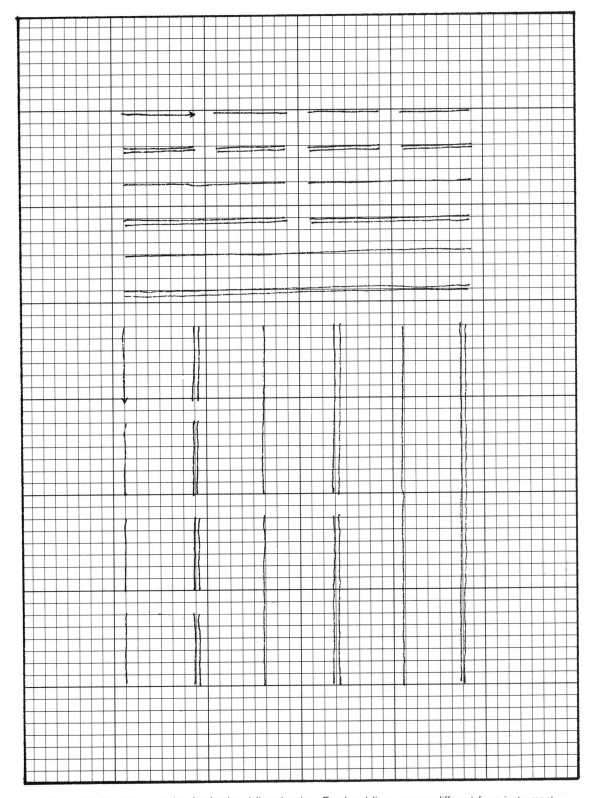

Figure 2-2. Practice exercise for freehand line drawing. Freehand lines appear different from instrument drawn lines. Do not be concerned about slight squiggles.

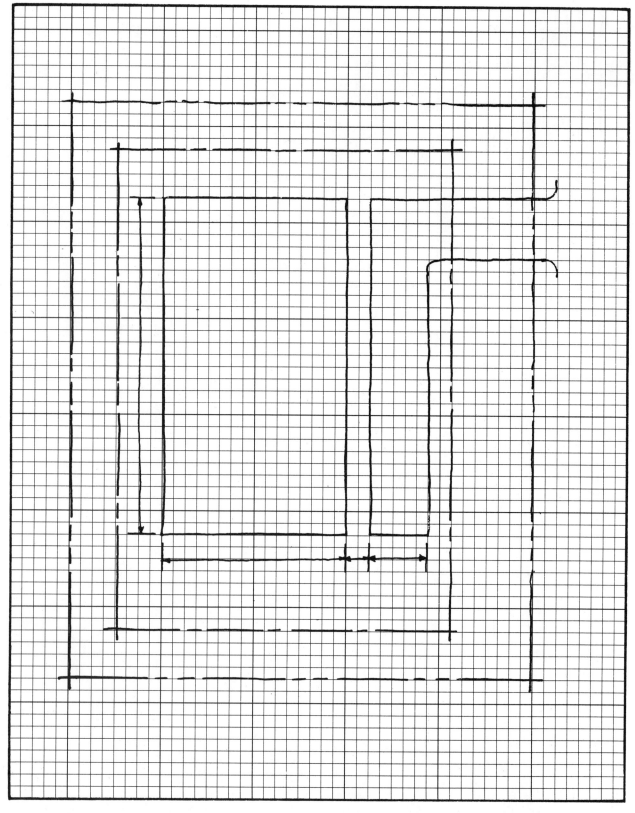

Figure 2-3. Freehand sketch of a site plan on graph paper. Note the use of different types of lines to indicate the various elements of the plan. Freehand sketching is commonly used during field site and building surveys. Graph paper is generally used.

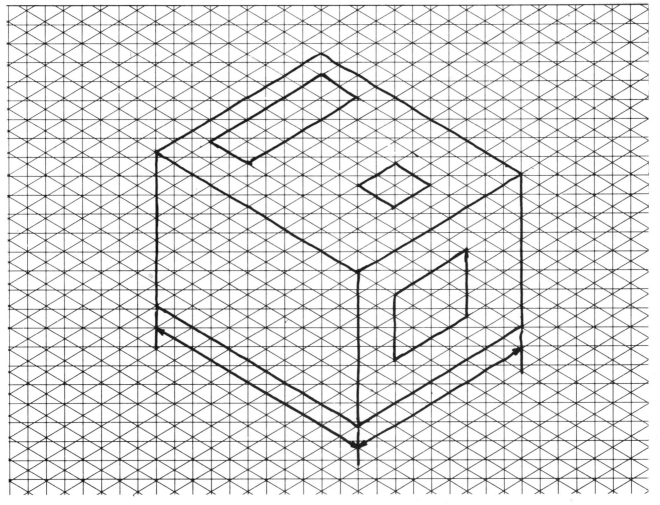

Figure 2-4. Isometric graph paper is used to make three-dimensional drawings of buildings and equipment.

30° line in either direction. All that is necessary is to follow the lines to develop any type of three dimensional (isometric) drawing (Fig. 2-4, Fig. 2-5).

Another valuable aid is what is known as the underlay drawing guide (chart). These charts are available in plastic sheets and opaque panels and can be used to develop one-, two-, and three-point perspectives, interior, exterior, isometric, and axonometric drawings. These charts are slipped under any type of tracing paper, vellum or polyester drafting media for freehand sketching or for hard line drawings (Fig. 2-6, Fig. 2-7).

Anyone involved in building construction drafting should develop freehand drawing skills. First, learn the technique and then practice until sufficiently proficient. Next, use the appropriate mechanical aids such as graph paper and underlay drawing guides. The quadrille and isometric drawing sheets will satisfy most needs in building construction drafting.

b. Hard Line (Instrument) Drafting

A hard line is one that is drawn with the aid of mechanical devices such as a ruling straightedge, triangle, T-square, french curve, etc. All building construction (working) drawings are done in hard line. Preliminary drawings and some-

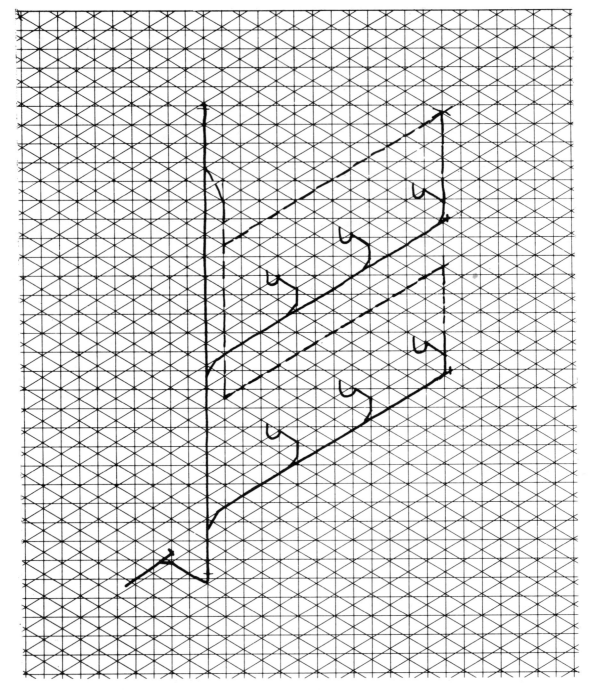

Figure 2-5. Plumbing riser diagrams can be easily and quickly sketched on isometric graph paper.

times even schematic drawings are also hard lined. Hard-line drawings provide the precision required to illustrate the design intent. The more precise the drawing, the less the probability of error during construction.

Precision line work is only one of the important requirements of quality drawings. Accurate scaling and dimensioning are other requirements, and legible and neat lettering are also very important. Hard line drafting is performed in much the same manner as freehand work. The right-handed drafter draws the lines from left to right. The only difference is that all lines are made with mechanical aids. This naturally simplifies drafting, once a straightedge is set in place, a line can be drawn along its edge.

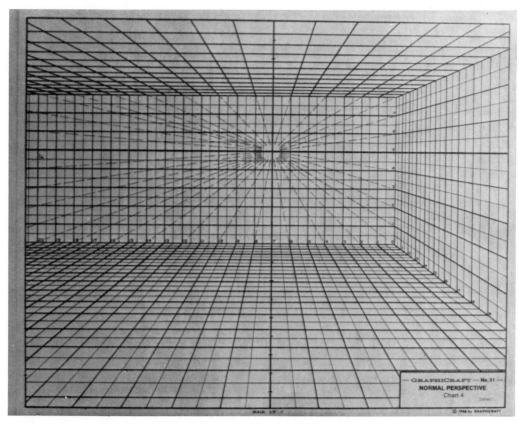

Figure 2-6. Plastic underlay drawing charts (guides) may be used as a freehand or instrument drafting aid. Place it under any transparent drawing medium and follow the lines on the chart.

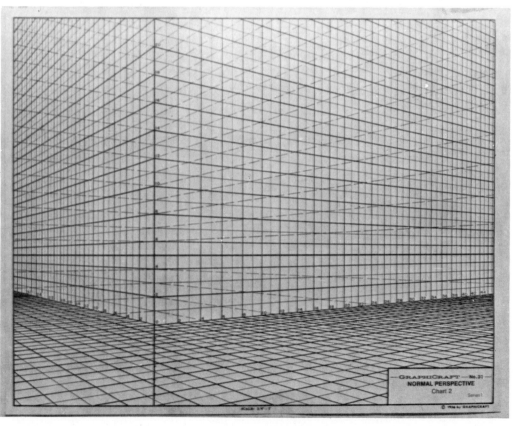

Figure 2-7. Normal perspective chart for exterior views. Plastic underlay drawing guides are available in various types and sizes from 11 × 14-inch to 22 × 28-inch sheets.

When drawing pencils and lead holders are used, they should be rotated in the fingers to produce a uniform point wear and hence, a uniform line width. If these devices are not rotated, the lead wears on one surface and the line width increases as the lead wears flat on one surface. Pencils and lead holders are held at a slight angle in the direction of the draw. Mechanical pencils and technical pens are held perpendicular and are not rotated since the line thickness does not change with use. Any lead that must be pointed must be rotated.

In hard line work, as well as in freehand drafting, the weight (thickness) of the line is essential to differentiate between the different elements of the building. Broken and segmented lines are also used to identify the various components, materials, equipment, dimensions, and section cuts. Quality drawings require the use of accepted line techniques for easy readability of the diagrammatical presentation. The weight of the line indicates its importance.

Building construction drawings are made to a reduced scale in order to show the structure on a manageable sized drawing. The amount of reduction (scale of the drawing) depend on several factors:

1. The scale should be large enough to permit all necessary information to be included. Where additional details are required, large-scale blowups are made. Working drawings are usually developed at either ⅛ inch or ¼.

2. The sheet should be of a size that is easily handled in the field during construction. Working drawings are generally not smaller than 24 x 32 inches and not larger than 32 x 48 inches.

Two types of reduction scales are used for building construction drawings, i. e., architect's open divided scales, and engineer's fully divided scales. Architect's scales are used for building and equipment design, and engineer's scales are used for site work. Both scales are designed to simplify drafting. For example, on the scale face where ⅛″ = 1′-0″ as identified by the fraction ⅛ on one end of the face adjacent to the ''0'', each mark on the scale to the right of the ''0'' represents one foot. On this scale, usually only every fourth mark is identified by the numbers 4, 8, 12, etc. To the left of ''0'' are six lines and each line represents 2 inches. To lay out a line of any length, start at the next lower number and read past the ''0'' mark for the fraction of the foot. To read 20′-6″, place the number 20 on the point and mark the end of that line at the third subdivision to the left of ''0'' to provide the correct length (see Fig. 5-4).

Initial building and equipment layout work is generally done with very lightweight lines using 2H or 4H lead, and these are later completed with the appropriate weighted lines as required for that particular segment of work. More experienced drafters and designers may use some lightweight lines to establish perpendicular boundaries and heavier lines directly in between. In some instances, a blue pencil is used for the preliminary outline since these blue lines do not reproduce when prints are made from the drawings, and do not have to be erased for that reason and therefore are often used for preliminary layout work and guidelines for lettering.

When using scales to lay out buildings and equipment, all dimension marks should be made without moving the scale along the drawing for each additional mark. For example, to establish the location of four columns spaced 20 feet apart, set the scale and make a mark at every 20-foot location starting from 0 through 80 feet. This technique eliminates the probability of cumulative errors that could occur when the scale is moved for every 20-foot mark.

The most important aspect of hard line drawings is accuracy in scaling. Even though the drawings are fully dimensioned, they need to be accurately scaled.

Second, appropriate line weights must be used to differentiate among the various elements of the building and equipment. Finally, line work must be sharp and neat.

2. Lettering

In building construction drafting and design neat, legible lettering is as essential as the hard-line drawing. Nothing detracts more severely from a good hard-line drawing than sloppy lettering. An additional requirement for pleasant appearing lettering is uniformity. Uniformity in height, proportion, inclination, weight of lines, spacing of letters, and spacing of words ensures a pleasing appearance.

Single stroke Commercial Gothic, in which the stems of the letters are made with a single freehand stroke, are very legible and can be rapidly executed. Remember, *time is money.* Drafters must be able to make both vertical and inclined letters with skill and speed. To do this requires diligence and continued practice, with careful attention to details.

Letters composed into words are not placed at uniform distances apart, but are spaced so that the words appear to be uniform. This is done by making the areas of white space between the letters appear equal. Anyone can learn to letter well if he is persistent and intelligent in his efforts. This includes capital and lower case letters.

Most architects and engineers develop their own personal style after they have learned the basics of good lettering techniques. Check any set of good quality building construction drawings and it soon becomes obvious that while the lettering may be similar in size and weight, there will be a noticable difference among the various styles on the sheets as they were most likely drawn by different people. While there may be this difference in the exact construction of the lettering as made by various individuals, the lettering will most likely be neat, legible, and appear uniform.

a. Freehand

At one time the term "draftsman's Gothic" applied to the letter style in general use by drafters. It may also be identified as "Commercial" Gothic, or "News" Gothic. The general construction of the letters and numerals are similar for all Gothic types. Basically, the so-called single stroke Gothic letters are easy to draw using a combination of single line straight and curved strokes, without any fancy frills. The primary intent of lettering on building construction drawings is legibility, and the primary goal of drafters and designers should be speed of execution (Fig. 2-8).

Today, many drafters merely copy the style they are exposed to without regard to the origin and most frequently end up making Gothic letters. Vertical single stroke lettering is easy to learn and with practice can be executed legibly and rapidly. Most strokes are natural and do not require memorization.

Gothic lettering consists of single stroke straight and curved line combinations. All vertical strokes are made from the top down and all horizontal strikes are made from left to right. Left handers may experience some difficulty with the order of strokes but the basic philosophy can be utilized. Left handers should experiment with the order of the strokes until letters can be constructed to their own particular style of writing. Vertical lines are made with a finger movement and horizontal lines with the hand pivoting at the wrist. Slanted strokes for letters such as A, V, W, X, etc., are also made from the top down.

ABCDEFGHIJKL MNOPQRSTUV WXYZ&

Figure 2-8. Drafters Gothic lettering. Note the use of various sized circles indicated to form the curved segments of certain letters. (Reprinted from Speedball Textbook, Property of Hunt Manufacturing Company)

There are several concepts on how curved letters should be constructed. Formal construction uses two or more discontinuous strokes for letters such as O, C, Q, G, etc., but in actual practice these letters are made with a single stroke, except for the tails on G and Q which require a second stroke. Figure 2-8 shows that the curved portion of the letters requiring curved strokes are essentially portions of circles. While some forms of Gothic lettering may use a somewhat oval shape in place of the circle, the difference is not all that significant. Most drafters, designers, architects, and engineers will eventually develop their own style to construct lettering with speed.

What follows is not intended to be a primer on perfect letter construction, but instead is a guide to practical lettering techniques that can be developed by anyone interested in pursuing a career in building construction drafting, and that requires legibility and speed, not perfect letter construction. Figures 2-13 and 2-14 illustrate the concept of single-stroke Gothic lettering.

Why is there so much emphasis on freehand lettering when there are several excellent mechanical lettering techniques that produce superior construction? Because most lettering on building drawings is freehand! In some offices all lettering is freehand, while in others the title block, column line designation, and similar notations are made by mechanical means.

Figures 2-9, 2-10, and 2-11 are practice lessons for developing finger and hand dexterity. Remember, vertical and slanted strokes are made with finger movement and horizontal strokes are made with the hand pivoting at the wrist. The use of cross-section graph paper is recommended for this practice for several reasons: first, the lines on the paper can be used as guides for all straight strokes; second, the height-to-width relationship can be easily determined by using the appropriate number of squares; and third, different sized letters can be practiced by using the horizontal lines to establish the height of the letters. Letter construction should be practiced until the proper concept of strokes are fully understood and the letters can be constructed without reference to the practice sheets.

After the basic fundamentals of good lettering construction are learned, diligent practice is required to develop legibility and speed. There is no one standard type of lettering used in actual practice; just check some of the illustrations in this book and it soon becomes obvious that lettering is a personal technique. While all experienced drafters, designers, engineers, and architects can letter

Figure 2-9. Elementary practice exercise for vertical and horizontal stroke Gothic lettering. (Reprinted from Speedball Textbook, Property of Hunt Manufacturing Company)

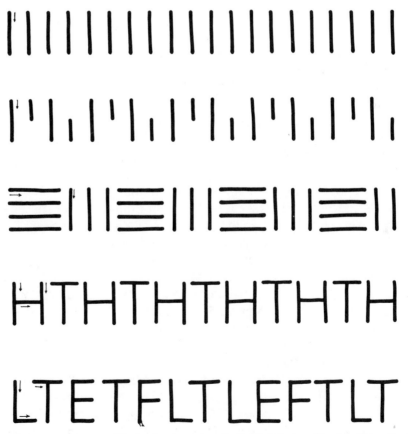

Figure 2-10. Elementary practice exercise for slanted stroke Gothic lettering. (Reprinted from Speedball Textbook, Property of Hunt Manufacturing Company)

Figure 2-11. Elementary practice exercise for circular and curved stroke Gothic lettering. (Reprinted from Speedball Textbook, Property of Hunt Manufacturing Company)

well, not all of their styles are exactly the same. Do not dismay, develop a technique that produces legible lettering that can be performed rapidly.

The next step is composing letters into words. The spacing of the letters can be either uniform (mechanical) or nonuniform (optical). In mechanical spacing some letters *appear* too far apart while others *appear* too close. In optical spacing some letters are placed closer together than others to give the *appearance* of uniformly spaced letters. When the areas of white space between the various letters appear equal, that spacing provides a more pleasing effect (Fig. 2-12).

In actual practice, capital letters are usually made of uniform height using a pair of parallel guidelines. These lines may be made using a lettering guide or drawn with the aid of a parallel straightedge and they are usually drawn with a hard lead such as No. 4 or 6 depending upon the preference of the drafter. To draw uniformly spaced lines with a straightedge, first lay out the spacing with an architect's scale, then draw the parallel lines. Some of the more experienced people will lay out the spacing by ''eye'', that is, draw the lines at a spacing that appears to be at an equal distance apart. This is done to save time. When using parallel guidelines, always draw up to and on the lines. For example, the top horizontal stroke of letters such as ''E'' is on the top line and the bottom stroke is on the bottom line. This produces uniformly sized letters and provides a pleasing appearance. In addition to the horizontal guidelines, many drafters also draw a perpendicular line to establish vertical alignment. Figure 2-13 is good example

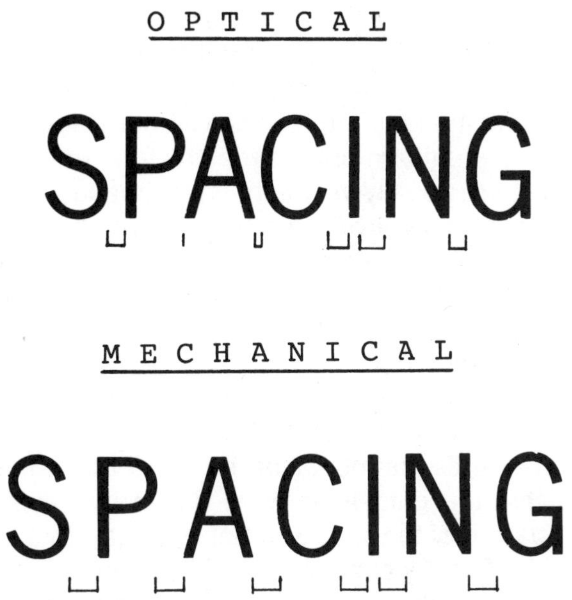

O P T I C A L

SPACING

M E C H A N I C A L

SPACING

Figure 2-12. Optical spacing provides a more pleasing appearance than mechanical spacing. The technique should be practiced until lettering can be performed without concern for spacing.

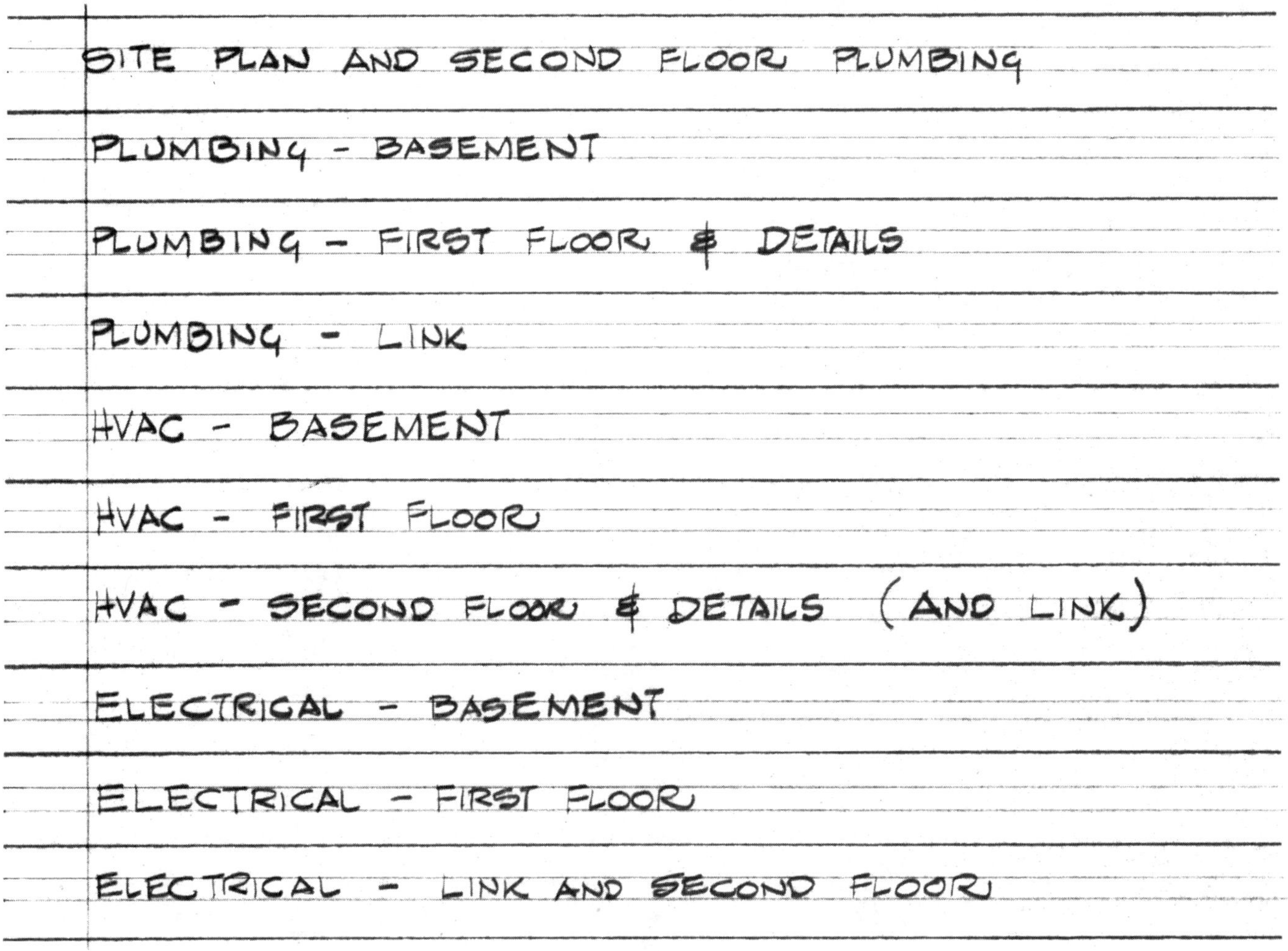

SITE PLAN AND SECOND FLOOR PLUMBING

PLUMBING - BASEMENT

PLUMBING - FIRST FLOOR & DETAILS

PLUMBING - LINK

HVAC - BASEMENT

HVAC - FIRST FLOOR

HVAC - SECOND FLOOR & DETAILS (AND LINK)

ELECTRICAL - BASEMENT

ELECTRICAL - FIRST FLOOR

ELECTRICAL - LINK AND SECOND FLOOR

Figure 2-13. Good lettering requires vertical alignment and horizontal guidelines. These guidelines are lightly drawn, barely perceptible, and do not detract from the lettering. The heavy lines indicate division between lines of lettering in the table.

of vertical alignment whereas Figure 2-15 illustrates what can happen without the perpendicular line. In cases where speed is most essential, some experienced architects and engineers use the straightedge device to establish the bottom of the letters and the lettering is done without guidelines. In these cases the base of all letters are uniform but the tops may be uneven.

Lettering guidelines are usually very light lines that can be seen on the drawing, reproduce on prints as faint background lines, and do not detract from or interfere with the lettering. These guidelines may be made with blue lead, and are visible on the drawing to serve the intended purpose but will not reproduce on the prints.

Figure 2-13 is an example of good vertical alignment and legible lettering. The horizontal guidelines are faintly visible and do not detract from nor interfere with the lettering. Also note the faint vertical line that provides vertical alignment. Careful examination reveals that not all of the horizontal guidelines are equally spaced, indicating that the guidelines were drawn by "eye" along a parallel straightedge. The spacing of the lines was not laid out with a scale. This method works well if the drafter is very experienced. Even though not all the letters are exactly the same size, the lettering is well articulated and easy to read.

Figure 2-14 illustrates the use of different sized letters to indicate the importance of the various portions of the "Electrical Symbol List." The main heading is ³⁄₁₆ inch high, the subheading is ⅛ inch high, and the body of the text is just

ELECTRICAL SYMBOL LIST

SYMBOL	DISCRIPTION
▭	FLUORESCENT LIGHTING FIXTURE
◯	INCANDESCENT LIGHTING FIXTURE - CEILING MTD.
◯┤	INCANDESCENT LIGHTING FIXTURE - WALL MTD.
"A"	INDICATES FIXTURE TYPE
100	INDICATES FIXTURE WATTAGE
S	SINGLE POLE SWITCH
S₃	LOCAL THREE WAY SWITCH
a	INDICATES SWITCH CONTROL
SM	MOTOR SWITCH - WITH THERMAL OVERLOAD PROTECTION
⊠	MAGNETIC STARTER
◯	MOTOR OUTLET
⬚┤	SAFETY SWITCH
▬	LIGHTING PANEL
▨	POWER PANEL

Figure 2-14. Different sized letters improve the appearance and should be used whenever headings and subheadings are required.

ROOM N°	DESCRIPTION
1	ELEC. RM
2	READY RM
3	READY RM
4	READY RM
5	COMPUTER CENTER
6	FIELD ENGINEERING
7	STORAGE
8	KEY PUNCH
9	STORAGE
10	MECH. RM
11	OFFICE
12	OFFICE
13	OFFICE
14	CONF. RM.
15	CORRIDOR / SECTY

Figure 2-15. Note what happens when vertical alignment guidelines are not used.

over ¹⁄₁₆ inch high. This is a good plan for lettering where different sized letters are required and is an excellent illustration of a symbol list. Here again, note the use of guidelines as indicated by the faint background parallel lines.

While the drafter who constructed the illustration in Figure 2-15 did use parallel guidelines, he did not use a vertical line for vertical alignment, and this detracts from otherwise good lettering.

Figure 2-16 is an example of lettering used on an electrical detail. Horizontal and vertical guidelines were used for all lettering as evidenced by the uniformity. Lines of lettering that are a continuation of the line above are closer together than the lines that indicate separate items of equipment. In this case the drafter combined vertical lettering for the notes with inclined lettering for the title. While the lettering is well executed and legible, the combination of types of lettering is distracting.

Good drafters always use horizontal and vertical guidelines, different sized lettering to indicate importance, and always provide greater space between different line items than between continued lines. Good drafters learn to letter legibly and rapidly.

b. Mechanical Lettering

Mechanical lettering, as addressed here, is done with the aid of a mechanical device and is generally used in title blocks and may be used for main titles on the drawing, and sometimes are even used for column line identification. Mechanical lettering is slower than freehand lettering and for that reason is not used for general drafting work on building construction drawings. There are, however, very specific and definite uses for mechanical lettering. Well-executed mechanical lettering by experienced drafters can provide near perfect letters. Three different methods of mechanical lettering are described here; lettering template, mechanical scriber, and dry-type transfer lettering.

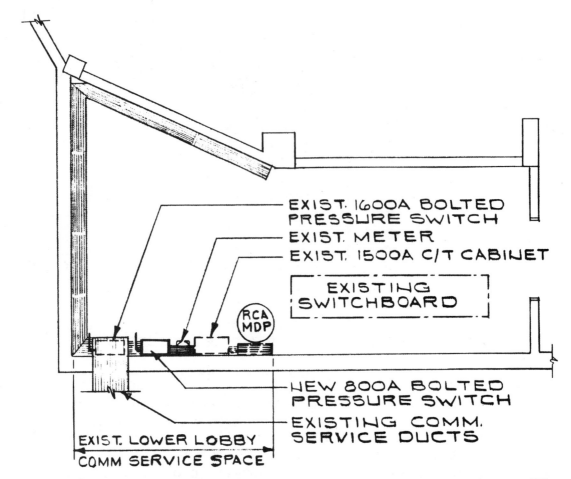

EXIST. 1600A BOLTED PRESSURE SWITCH
EXIST. METER
EXIST. 1500A C/T CABINET
EXISTING SWITCHBOARD
RCA MDP
NEW 800A BOLTED PRESSURE SWITCH
EXISTING COMM. SERVICE DUCTS
EXIST. LOWER LOBBY
COMM SERVICE SPACE

MAIN ELECTRIC EQUIPMENT ROOM
B-2 LEVEL
SCALE 1/8"=1'-0"

Figure 2-16. Good lettering practice improves the appearance of detail drawings. In this case vertical lettering was used for the text and inclined (slanted) lettering was used for the title.

1. Lettering Template

Lettering templates are made of rigid, transparent clear to colored (orange or green) plastic in thicknesses of 0.20 to 0.60 inch. These cut-through lettering guides are available in a wide variety of vertical and inclined letter styles, upper case and lower case letters, and in a number of different size letters. They are available in the type that has only one size, and the more common type that has three sizes, usually ⅛, ³⁄₁₆ and ¼ inch on one template. The latter is more frequently used by drafters because the three most commonly used sized letters and numerals are combined in one template.

Lettering templates are easy to use with either lead or ink. For best results they should be used with a T-square or a parallel straightedge to establish uniform alignment. Some have equally spaced guidelines that are used to ensure equal spacing between the letters and numbers. Some have recessed design to prevent ink smears; others have elevated plastic rails for use with ink. Edge strips may be purchased, cut to fit and installed on flat lettering guides to prevent ink smears.

To use a lettering template place it along the parallel straightedge and slide it back and forth as needed to properly space the individual letters. The result, when properly used is uniformly formed letters and numbers and a professional appearance. These lettering templates may also be used for practice to develop a feel for correctly formed letters and uniformly composed words.

2. Mechanical Scriber

Another method of mechanical lettering uses a scriber and scriber template. The templates have engraved grooves to form the letters. The template letters are not cut through as in lettering templates. The scriber is simpler to use and the lettering can be made perfect. After the scriber is set in place on the template, all the drafter has to do is trace the grooved letters with the tracing pin. The pen reproduces the letter in an easy to see location above the template. A light, delicate touch reproduces perfect letters each time.

The Koh-I-Noor Rapidograph® controlled lettering set comes complete with 12 different sized scriber template guides for making letters from .060 to .500 inch high and 10 pen point sizes from 4 × 0 to No. 6, with a line width to letter height ratio of 1 to 10. Different sized points are recommended for use with different sized letter templates in order to reproduce appropriately weighted (width) letters (Fig. 2-17).

The pen threads securely into the pen receiver on the scriber. The scriber is

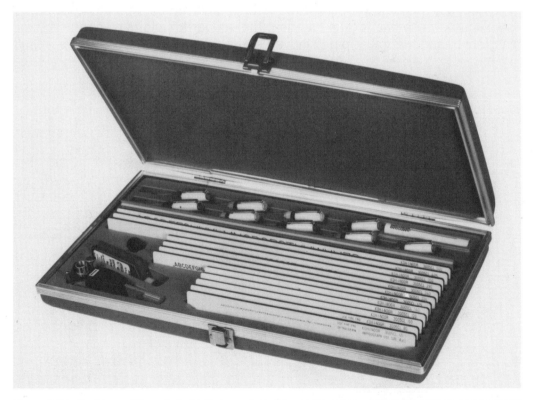

Figure 2-17. Rapidograph® controlled lettering set contains scribe guides, adjustable scriber, and technical pens of various sizes. (Courtesy of Koh-I-Noor Rapidograph, Inc.)

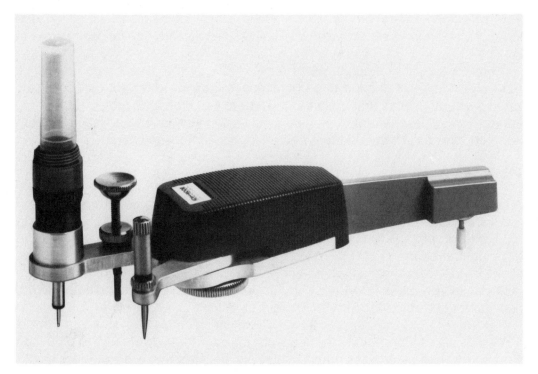

Figure 2-18. Scriber arm with technical pen in place. (Courtesy of Koh-I-Noor Rapidograph, Inc.)

fully adjustable to provide lettering slanted to any degree from the vertical to 22.5° (Fig. 2-18).

The scriber template with engraved letters is placed along a straightedge to establish the base line for the lettering, the scriber tail pin is placed in the tail pin groove, and the tracer pin is moved in the grooves of the guide. The result

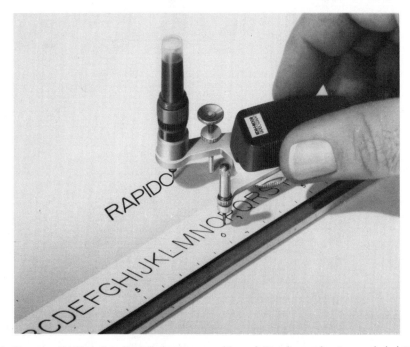

Figure 2-19. To use, set tail pin into the tail pin groove and trace letters by moving tracer pin in letter grooves to form the letters. A light, delicate touch is all that is necessary to produce neat, uniform lettering. (Courtesy of Koh-I-Noor Rapidograph, Inc.)

is neat, uniform lettering that adds a professional quality to the drawings. Anyone with a little practice can become proficient in the use of scriber lettering.

Controlled lettering sets are available with 3, 6, 9 or 12 different sized pen points and scriber guides as shown in Figure 2-17.

3. Transfer Lettering

Dry-type transfer lettering is also used on building drawings. The letters are durable nitrocellulose ink produced on transparent polyester carrier sheets. This heat-resistant transfer lettering is ideal for drafting room applications. The nitrocellulose ink transfers quickly and easily to polyester drafting film, vellum, paper, and cloth. Convenient 8¼ by 11¼ inch sheets provide easy maneuverability on the drawing board and permit file folder storage.

Chartpak® offers 295 different font styles in point sizes up to 180. Each set contains a wide selection of the most frequently used letters, number, and punc-

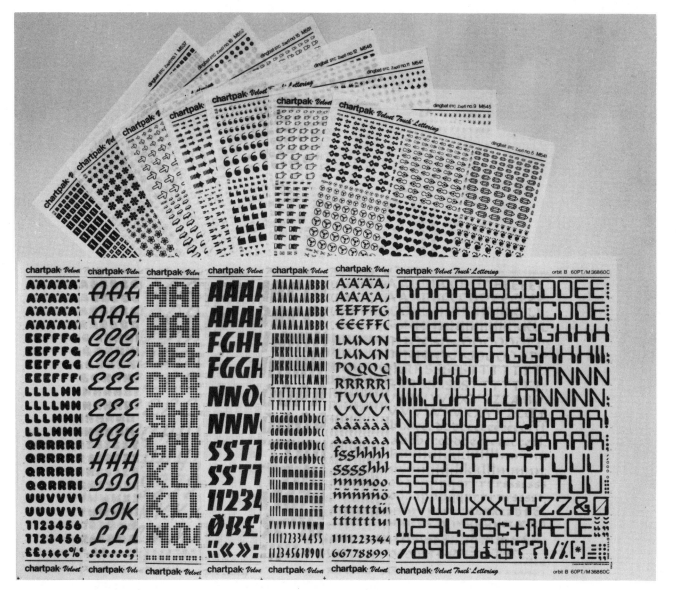

Figure 2-20. Sample assortment of some dry-type transfer letters, numbers, and symbols available. (Courtesy of Chartpak)

tuation marks. Careful research determined the most often used vowels, consonants, and punctuation marks. A wide variety of commonly used symbols are also available. A simple three-step operation is all that is necessary to transfer a letter from the carrier sheet to the drawing (Fig. 2-21).

- Carefully position the letter to be transferred.
- Rub over the entire area of the letter with the point end of the burnisher (a tool specifically designed for dry-type transfer lettering), a pencil or a ball point pen. Be sure to rub over all areas of the letter.
- Remove the carrier sheet by carefully lifting it by one corner while holding the sheet in place to check that the letter has been completely transferred. For maximum adhesion, place the backing sheet (or similar tissue paper) over the letters transferred and rub again with the ''bone'' end of the burnisher (Fig. 2-22).

APPLICATION:

Transferring Velvet Touch and Heat Resistant transfer lettering is quick and easy. Follow directions pictured below. In the event of error, corrections can be made easily with a pencil eraser, rubber cement pickup, or by lifting the transfer with transparent or masking tape. Where lettering will be exposed to abrasion or weather, it should be sprayed with Chartpak Clear Spray or Chartpak Workable Matte Fixative.

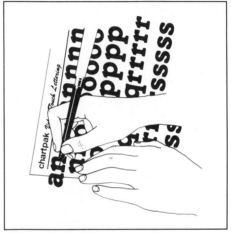

2. Rub over letter with the Chartpak burnisher - be sure to cover all areas and fine lines.

1. Position letter.

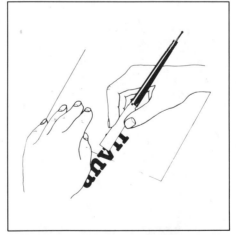

3. Remove sheet by carefully lifting from corner. For maximum adhesion, place backing sheet over letter and rub again with the white flat end of the Chartpak burnisher.

Figure 2-21. Dry-type transfer lettering application instruction. (Courtesy of Chartpak)

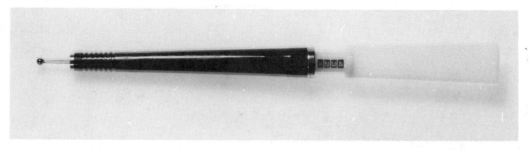

Figure 2-22. The point end of the burnisher is used to transfer letters, symbols, etc., from the carrier sheet to the drawing and the "bone" end to "set" lettering. (Courtesy of Chartpak)

There are several advantages to this method of lettering: it is a quick and easy method for producing sharp and clear letters; letters are precisely formed; a complete range of letter types and sizes are readily available on single sheets; in the event of error, corrections can be made easily with a pencil eraser, rubber cement pickup, or by lifting the transfer with transparent or masking tape; lettering will not creep or blister when run repeatedly through a diazo machine.

Each of the three mechanical lettering methods has its own unique advantage, depending upon application and circumstance. In most professional offices, all three are used. Drafters should know how to use each method and develop sufficient proficiency to produce neat, professional quality lettering with speed. However, since the major portion of all building construction drafting is freehand, that method requires the most attention and practice in order to be able to letter legibly and rapidly.

Chapter 3
Drafting Mediums and Print Making

Drafting mediums discussed here include those materials used in building construction drafting. They include sketching and tracing paper, vellum, polyester film, intermediate paper and film, and print paper. Initially buildings are drawn on transparent paper, vellum or film, which can be reproduced on intermediates (often referred to as duplicate tracings or second originals, and sometimes as sepias). Sepias are characterized by a sepia image on off-white paper base, and prints. The drawing is a means to an end, not the end in and of itself. When a drawing is completed it is the only copy, therefore copies must be made. It is not unusual to make several copies during the various design phases since many copies are generally required for bidding and construction purposes.

The drafting material selected must withstand repeated handling, multiple erasures without harm to the drafting surface, and numerous passes through the print machine and must reproduce well. The first step is to make the drawing on transparent material so that it can be reproduced. Frequently, intermediates are required to expedite drafting or for record purposes. Finally, prints are made for review of the design, coordination, estimating, bidding, and for use at the construction site.

1. DRAFTING MEDIUMS

A wide variety of drafting mediums are marketed, but only a few different types are used in building construction drafting. They all have one feature in common—transparency. The degree of transparency varies with the type of medium and to some extent by the manufacturer. For maximum utilization and effectiveness, the medium used must be suited for the purpose. For example, not all mediums are suited for use with ink, graphite, or polymer lead. Also, not all are equally durable, erasable, reproducible, resistant to moisture, and able to maintain good dimensional stability. Some are more general purpose and may be used with ink, graphite, or polymer lead, whereas others are specifically recommended to be used only with either ink, graphite, or polymer lead.

a. Tracing Paper

In the past, pencil drawings were made on off-white drawing paper and then *traced* frequently with ink, on tracing paper or cloth, (usually linen), placed over the drawing. The persons doing the tracing were called "tracers" and the paper used for this purpose was known as tracing paper. Today, drawings are made directly on the tracing paper, while the use has changed, the name remains.

Tracing Papers

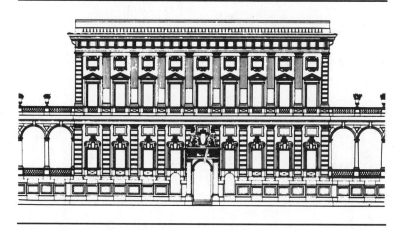

Architectural rendering of Doria Tursi Palace in Genoa, Italy. Circa 16th century.

Figure 3-1. 15-pound, 100% rag, natural tracing paper. (Courtesy of Ozalid Corporation)

Tracing paper is a thin, transparent paper on which drawings are made and can be reproduced by the diazo or similar process. There are several types of tracing paper; 100% rag, natural tracing paper; 100% rag, natural paper transparentized with resins; and 100% rag, natural paper polymerically transparentized. These types are divided into two general groups—tracing paper and vellum.

When the 100% rag-base paper is transparentized by a process which fills all interstices of the paper with a water-clear inert resin, the result is superior translucency and durability.

Tracing paper is available in various grades and is the most economical drawing medium currently marketed. It is ideal for the student, can be used with ink and graphite lead, and is used in some professional offices, depending upon circumstances.

A fine tracing paper is 100% rag, natural paper and offers the following features:

- Very white in appearance and has good light reflectance yielding high contrast ratios well suited for top-lit reproductions. Opacity is equal to or less than 70%.
- Excellent quality for inking with technical pens.
- Easy erasure of lead lines; erasure of ink lines is more difficult.
- Quite durable, will withstand considerable handling.
  Good reproduction properties, well suited for diazo reproduction.

Tracing paper is available in sheets ranging in size for 8½ × 11 inches and up, depending upon the manufacturer, and in various width rolls, generally 20 yards long. Sheets are made with typical title blocks designed to suit office requirements.

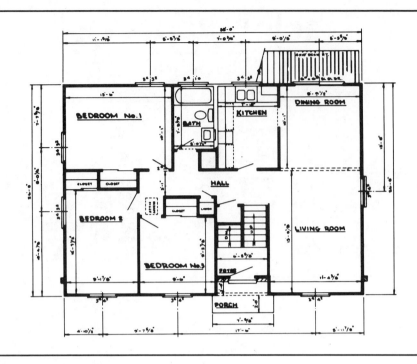

Figure 3-2. 16-pound, 100% rag, polymerically transparentized vellum. (Courtesy of Ozalid Corporation)

b. Vellum

Vellum is 100% rag paper thoroughly permeated with a transparentizer, which is permanent, nonyellowing, nontransferring, nonfracturing (from creases and dog-earing), and, in erased areas, nonghosting. Vellums are designed to accept both ink and graphite lead and to withstand multiple erasures without harm to the drafting surface.

Vellums are available in various grades, several different weights, and are slightly more expensive than tracing paper. It is more frequently used in professional offices than tracing paper, because of its superior features:

- Good transparency, with a high average percentage of light transmittance.
- Excellent ink line quality with all paper inks.
- Good graphite line properties. The smooth surface finish is highly resistant to marring and *ghosting* even after repeated erasures of graphite lines.
- Highly resistant to moisture and maintains good dimensional stability and permanence.
- Well suited for diazo reproduction.

Vellum is available in sheets, single and padded, and in rolls in the same general sizes as described above for tracing paper. Sheets may be purchased with border lines, and standard and custom designed title blocks.

c. Polyester Film

Polyester drafting film has a higher initial cost than paper or vellum, but manufacturers and users alike agree that superior results can be achieved with film in shorter time and with less effort than either paper or vellum. Film is generally used in professional offices where its higher cost is offset by its distinct advantages. Its primary advantage is that it is virtually indestructible. It also has the ability to maintain upmost line precision and highest dimensional stability for even the most demanding reproduction requirements.

Since the primary function of a drafting room is to communicate design information within the limits of time and budget, yet maintain quality for better understanding, polyester film can be tailored to meet these challenges. Drafting films are made by coating polyester film with a super-smooth drafting matte to provide an ideal medium for architectural and engineering drawings. The matte accepts lead and ink and withstands multiple erasures without affecting draftability. The transparent polyester film base is tough, dimensionally stable, and will not tear or dog-ear. Drawings on film are excellent masters for all methods of reproduction.

Polyester drafting film is available with matte on one or both sides and in a variety of sheet and roll sizes. Drafting film is also superior to any other medium in archival quality; it will not crack or yellow with age. It can withstand the abuse of repeated handling since it will not rip, stretch or stain, and unlike paper, there is seldom a need to throw away any piece of film.

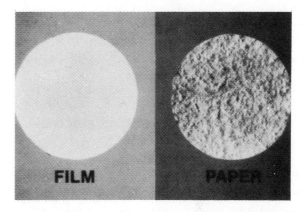

Figure 3-3. Magnified photo compares controlled, even, nonfibrous surface of Cronar® polyester film with fibrous surface of paper. Cronar® is a registered trademark of the Du Pont Company. (Courtesy of the Du Pont Company)

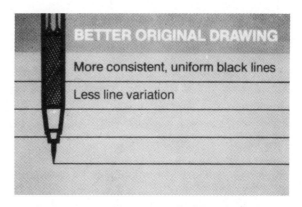

Figure 3-4. Plastic lead on film produces a better original than graphite on vellum. (Courtesy of the Du Pont Company)

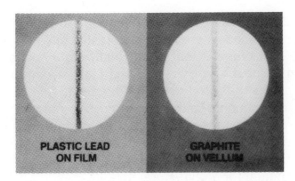

Figure 3-5. Magnified photo contrasts pencil line work on film with that on vellum. (Courtesy of the Du Pont Company)

Figure 3-6. Jewel point technical pens with ink specially formulated for films produce superior uniform line quality. (Courtesy of the Du Pont Company)

The most used film systems, with plastic lead, produce a better original than graphite on vellum. Lines are more consistent, more uniform, and blacker. There is less line variation among drafters. On film, the lines are more uniform as less pressure is needed to put them down because of the film's smooth surface. In contrast, a line on vellum is put down by the fibers snipping off bits of graphite, which results in less uniformity.

Because film has a controlled even surface compared to the fibrous less-even surface of paper, laying down the line will feel different. The technique that works well on paper will not necessarily work well with film. For best results, follow these suggestion:

1. Clean the drafting film surface with a dry cloth before starting drawing, and keep the surface clean.
2. Plastic lead is recommended over graphite for most applications.
3. Use a slightly blunted point and lighter pressure to avoid embossing the film surface.
4. Maintain constant wrist angle while drawing the entire length of the line.
5. Draw in one direction only, without retracing the line, to avoid two layers of plastic lead.
6. A firm, rather than pliable backing is best.

The following are plastic and combination lead suggested for use with Cronar® polyester film. (Cronar® is a registered trademark of the Du Pont Company)

- Eagle Filmograph
- Staedtler Mars Dynagraph
- Faber-Castell-Filmar
- Ruwe

There is a growing trend toward ink on film. This trend to ink is significant, quite changed during the last decade. The reasons?. . . refinement of technical pens, including jewel points . . . inks specially formulated for films . . . vinyl erasers . . . and improved film drafting surfaces. This combination of developments has made ink on film a fast-growing favorite, because the quality of the finished original is significantly improved over graphite on vellum. With ink, the line quality is superior, sharp, uniform, and continuous, and redrawn lines duplicate original lines.

For the crisp, accurate lines only ink produces, these techniques with drafting films are recommended:

- Use technical reservoir type pen with jewel point.
- Maintain clear drafting surface. Do not use pounce.
- Use a light touch, holding the pen perpendicular to the surface.
- Keep the pen capped, when not in use, to prevent ink from drying in the point feed mechanism.

Suggested supplies for inking:

Technical pens	K & E Leroy® Reservoir pen
	Koh-I-Noor—jewel point
Ink	Staedtler Mars
	Pelikan "T"
	Higgins Black Magic
	Koh-I-Noor 3071, 3072

Design changes are a drafting fact of life. Here is one way to accomplish them successfully on drafting film:

- Use nonabrasive vinyl erasers.
- Slightly moistened for plastic lead or ink work.
- Dry if graphite lead is used.
- Erase at right angle to the line.
- Do not use an electric eraser. The qualities of the drafting film surface makes this unnecessary.

Suggested erasers for film:
Faber-Castell Magic Rub 1954
Koh-I-Noor   286
Staedtler Mars 526-50
Dietzgen Film Rub 3329
K & E Tad No. 580600

d. Sketch Paper

Sketch paper is an 8-pound lightweight economical paper used by architects and engineers for idea sketching, preliminary detail work, and rough layout drawings. It is available in transparent white and canary paper, in rolls of several sizes from 12 to 42 inches wide and 50 yards long. The paper is receptive to ink, pencil, and felt-tip markers, and reproduces well. Felt markers and soft carbon pencils are most frequently used. The use of ball point pens and hard drafting lead is not recommended.

Sketch paper may be placed directly on drawing boards or drafting tables for rapid development of preliminary ideas, or it can be placed over drawings for sketching equipment layouts, space planning, design changes, etc. Since the material is transparent, several sheets may be overlayed for the development of several concepts. These sketches are then given to drafters as guides for instrument (hard-line) drafting.

Because of its light weight, sketch paper does not hold up well under erasures, and most frequently, unsuitable sketches are discarded. Appropriate sizes of the paper, as required, can be simply torn from the roll without aid of special equipment. The paper tends to adhere slightly to any surface without the need to be taped into place during use. Sometimes, however, it is taped into place for precise overlay control.

While larger width rolls may be easily "sawed" into smaller sizes as required, the recommendation is to purchase the size that best suits the anticipated use. The sketch paper is an excellent material for the student drafter to develop various plans, views, etc., and is essential for use in professional offices.

2. INTERMEDIATES

Intermediates, often referred to as duplicate tracings or second originals, and also known as sepia prints, are a necessary adjunct to the drafting process and permit the use of many time- and cost-saving techniques in the duplication and distribution of drawings (Fig. 3-8).

An excellent example is the duplication of basic floor plans drawn by architects and the intermediates are used by the various engineering disciplines to fill in their own specific design information. In this case, the basic building is drawn only once and the intermediates are used for HVAC, plumbing, fire protection, and electrical work. This technique not only saves drafting time but also provides an exact building background for all discipline drawings.

Intermediates also save many hours of drafting time during the design of multi-story buildings. These high-rise buildings frequently have numerous floors with identical perimeters, column locations and core sections. After the floor plan is drawn once, intermediates are made for each floor and the variables (HVAC, electrical, plumbing, fire protection and partition layouts) are drawn in.

Many architectural and engineering (A/E) firm contracts require the contractor to furnish the owner with what is called "as-built" drawings. The contractor is supplied a set of intermediates that show the building as designed and the contractor then corrects the intermediates to reflect any changes made during construction.

Many professional offices do not permit the original drawings to be removed from the office for any reason. In this case, intermediates are made for distribution as required.

There are two types of intermediates—paper and film. Paper intermediates yield an opaque diazo sepia image on an off-white translucent background. Depending upon the type of intermediate paper used, the background could have a slight blue or green tint or a sepia color. Paper intermediates are therefore frequently called sepias. Again, depending upon the type of paper used, the diazo image can be removed either chemically or mechanically. With a dry erasable sepia the image is removed with a slightly abrasive eraser. Other paper intermediates require a liquid eradicator. In either case, portions of the drawing can be removed and revisions may then be drawn on the sepia. Most paper intermediates have excellent drafting surfaces. Also, unwanted portions of a drawing can be cut out of the original or sepia, leaving a clear area on an intermediate print for drafting of the revised design. Intermediates can be used as masters for making sepias or prints on any diazo sensitized paper or film.

Film intermediates have all the same advantages over paper intermediates that film originals have over vellum and tracing paper. Film intermediates are more expensive than paper intermediates.

The photographic image on DuPont CROVEX® moist-erasable reproduction film is easily removed; merely moisten, erase, and redraw. The film is dimensionally stable, archival and has a drafting surface ideal for ink or lead. The image is black on clear transparent polyester film (Fig. 3-7).

Many drawing inks are satisfactory for use with original and intermediate films. Some have been specifically designed for the purpose, and these should be the first choice. High viscosity, quick drying inks work best. Check with the particular film manufacturer to obtain the recommended list of pens, ink, lead, and erasers.

Polyester and vellum pressure-sensitive applique sheets can be used with polyester films, and vellum applique sheets are generally used with tracing paper and vellum. Reverse printed applique sheets (applied to the underside of drawings) are recommended unless the drawings are to be microfilmed.

3. PRINTS AND PRINTMAKING

The term "blueprints" refers to prints made from drawings where the background is blue and the line work is white. In early days, dating back over a century, building construction drawings were reproduced as blueprints. This method of reproduction has largely disappeared but the term is still frequently used. Today, prints are the reverse of blueprints, that is, the background is white and the line work is dark, either blue or black, hence the name, blueline or blackline prints. The term blueprints is still frequently used in building construction work, particularly by oldtimers who used such prints, to refer to blueline and blackline as blueprints. Most frequently, the various types are called *prints*.

The most common method for making prints from transparencies is the diazo process. The Ozalid™ system is widely used in many professional offices and in reproduction service centers (Fig. 3-9).

The technology of the diazo process is simple when compared with the complex reactions that occur when reproducing copies with photographic or xerographic processes. With the diazo process, only two basic steps take place during the making of a print—exposure and development.

During the making of a diazo print, ultraviolet light passes through the translucent areas of the original (drawing), strikes the diazo compound, and destroys

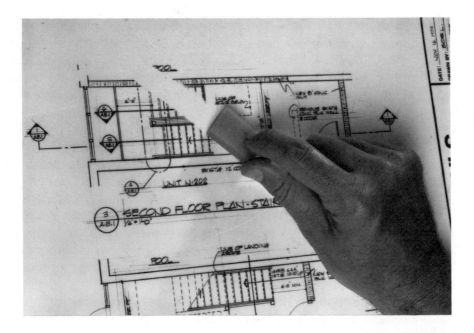

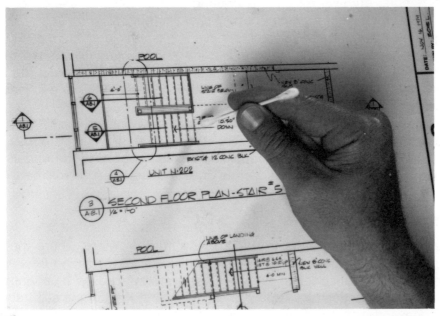

Figure 3-7. Photographic image on Crovex® wash-off film is moist-erasable. Moisten the image to be erased with a water-soaked cotton swab or other technique, let stand for 5 seconds. Use a vinyl, plastic, or imbibed eraser to lift unwanted lines, letters, or images right off the moistened film. Crovex® is a registered trademark of the Du Pont Company. (Courtesy of the Du Pont Company)

it. The opaque image area (line work and lettering) of the original blocks the ultraviolet light, thereby preventing exposure of the diazo compound. The unexposed diazo, or that area corresponding to the image area of the original, forms a dye during development.

The exposure of diazo materials involves a photochemical reaction, or one in which radiant energy is used to produce a chemical change. The radiant energy which produces the chemical change is ultraviolet light; the chemical change is the decomposition of the bright yellow diazo compound into a colorless, chemically inactive, decomposed product.

Intermediate Papers

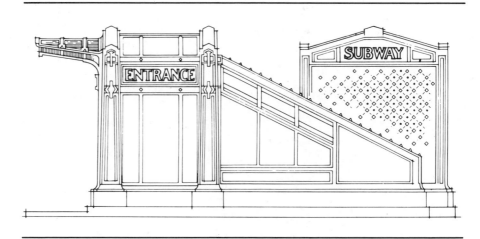

Rendering of original New York City subway station entrance. Circa 1902.

Figure 3-8. Intermediate paper—sepia. Features dry image, erasable using a slightly abrasive eraser. (Courtesy of Ozalid Corporation)

The diazo process is based on the chemical reaction that occurs when the unexposed diazo compound reacts with the couplers, or color forming, compound, in an alkaline (ammonia) environment to form a dye image. This chemical reaction is commonly referred to as coupling.

Every diazo coating, in addition to diazo and couplers, contains mild acids called stabilizers. The coupling reaction will not take place in the acid environment created by these stabilizers. In order for the coupling reaction to take place, the acid stabilizer must be neutralized by an alkali or base.

Ammonia, because of the ease with which it can be vaporized, the evenness with which ammonia vapors fill an open area such as a developer tank, and its low cost, it has been found to be the most effective and efficient substance for creating the alkaline environment needed to neutralize the stabilizing acid. Ammonia is commonly referred to by its chemical symbol, NH_3. Ammonia vapors react with and neutralize the stabilizing acid. An alkaline environment is created. The diazo and coupler then react with each other to form a dye image.

The simplicity of this reaction, which eliminates costly and inconvenient chemical developers, is a major advantage of the diazo process over other reproduction systems.

To see how to make a diazo print or diazo intermediate, follow a translucent original and a piece of diazo material through the path of a typical diazo machine. These are the events that take place during processing (Fig. 3-10):

1. The original is placed in contact with the sensitized material.
2. The original and the sensitized material are exposed to ultraviolet light.
3. The original is separated (either manually or automatically) from the exposed sensitized material.

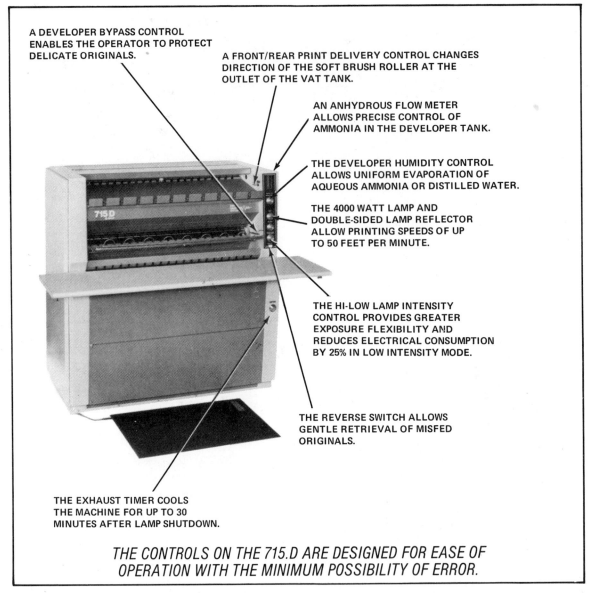

A DEVELOPER BYPASS CONTROL ENABLES THE OPERATOR TO PROTECT DELICATE ORIGINALS.

A FRONT/REAR PRINT DELIVERY CONTROL CHANGES DIRECTION OF THE SOFT BRUSH ROLLER AT THE OUTLET OF THE VAT TANK.

AN ANHYDROUS FLOW METER ALLOWS PRECISE CONTROL OF AMMONIA IN THE DEVELOPER TANK.

THE DEVELOPER HUMIDITY CONTROL ALLOWS UNIFORM EVAPORATION OF AQUEOUS AMMONIA OR DISTILLED WATER.

THE 4000 WATT LAMP AND DOUBLE-SIDED LAMP REFLECTOR ALLOW PRINTING SPEEDS OF UP TO 50 FEET PER MINUTE.

THE HI-LOW LAMP INTENSITY CONTROL PROVIDES GREATER EXPOSURE FLEXIBILITY AND REDUCES ELECTRICAL CONSUMPTION BY 25% IN LOW INTENSITY MODE.

THE REVERSE SWITCH ALLOWS GENTLE RETRIEVAL OF MISFED ORIGINALS.

THE EXHAUST TIMER COOLS THE MACHINE FOR UP TO 30 MINUTES AFTER LAMP SHUTDOWN.

THE CONTROLS ON THE 715.D ARE DESIGNED FOR EASE OF OPERATION WITH THE MINIMUM POSSIBILITY OF ERROR.

Figure 3-9. Ozalid™ 715.D diazoprinter. (Courtesy of Ozalid Corporation)

4. The exposed sensitized material is transported into the development chamber for saturation with ammonia vapors.

5. The developed diazo print (or intermediate) is delivered to the operator.

The diazo process can be used to make blackline and blueline prints, paper intermediates, and film intermediates.

Diazoprinters are available in sizes ranging from small tabletop units to large free-standing, fully automatic machines capable of handling prints up to 47½ inches in width. With the small tabletop units, the originals and the sensitized print material is first passed through the ultraviolet section, then the two are manually separated and the sensitized sheet is fed into the development chamber. With the large machines, this separation is performed automatically within the machine. The original and the sensitized material are fed into the machines; the original is separated from the print sheet and returned to the operator while

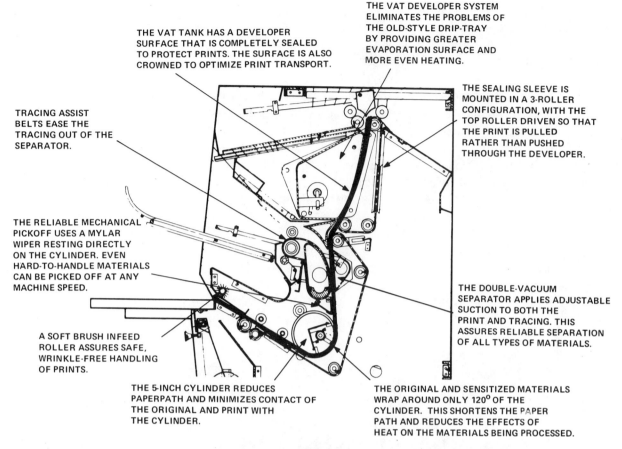

THE VAT DEVELOPER SYSTEM ELIMINATES THE PROBLEMS OF THE OLD-STYLE DRIP-TRAY BY PROVIDING GREATER EVAPORATION SURFACE AND MORE EVEN HEATING.

THE VAT TANK HAS A DEVELOPER SURFACE THAT IS COMPLETELY SEALED TO PROTECT PRINTS. THE SURFACE IS ALSO CROWNED TO OPTIMIZE PRINT TRANSPORT.

THE SEALING SLEEVE IS MOUNTED IN A 3-ROLLER CONFIGURATION, WITH THE TOP ROLLER DRIVEN SO THAT THE PRINT IS PULLED RATHER THAN PUSHED THROUGH THE DEVELOPER.

TRACING ASSIST BELTS EASE THE TRACING OUT OF THE SEPARATOR.

THE RELIABLE MECHANICAL PICKOFF USES A MYLAR WIPER RESTING DIRECTLY ON THE CYLINDER. EVEN HARD-TO-HANDLE MATERIALS CAN BE PICKED OFF AT ANY MACHINE SPEED.

THE DOUBLE-VACUUM SEPARATOR APPLIES ADJUSTABLE SUCTION TO BOTH THE PRINT AND TRACING. THIS ASSURES RELIABLE SEPARATION OF ALL TYPES OF MATERIALS.

A SOFT BRUSH INFEED ROLLER ASSURES SAFE, WRINKLE-FREE HANDLING OF PRINTS.

THE 5-INCH CYLINDER REDUCES PAPERPATH AND MINIMIZES CONTACT OF THE ORIGINAL AND PRINT WITH THE CYLINDER.

THE ORIGINAL AND SENSITIZED MATERIALS WRAP AROUND ONLY 120° OF THE CYLINDER. THIS SHORTENS THE PAPER PATH AND REDUCES THE EFFECTS OF HEAT ON THE MATERIALS BEING PROCESSED.

A Short, Straight Travel Path Assures High Productivity And Reliability

Figure 3-10. Travel path through the Ozalid™ 715.D diazoprinter. (Courtesy of Ozalid Corporation)

the print is passed through the development chamber. Machine speeds can be adjusted, depending upon the size and type of machine from 1½ to 40 feet per minute. A speed control dial governs the transport speed. The particular speed required depends upon the quality and type of the original and the type of sensitized print or intermediate material used.

In addition to the standard blueline and blackline on white background print paper, green stock and pink stock are available from some print stock manufacturers. With these colors a black image is made on the green and a blue image is made on the pink stock. These are used in those cases where special emphasis is required for important documents to distinguish them from the white background prints.

4. GOOD DRAFTING PRACTICE

Engineering and architectural drawings represent hundreds of hours of planning and drafting time. Therefore, it is important to use good drafting practice to protect this investment of time and effort. Good drafting practice includes selecting the appropriate drafting medium, using proper tools, protecting drawings during drafting, making maximum use of intermediates to expedite drafting, and properly storing originals and intermediates.

Select the drafting medium that best serves the purpose. Most residential and small commercial projects do not justify using polyester films. These projects are frequently one-time services that when completed do not require the retention of originals for follow-up work. In these cases paper or vellum satisfies the need.

On larger projects that include commercial, institutional and industrial buildings, polyester films are preferred. In these cases the originals may be retained for extended periods. The A/E firm is frequently retained for follow-up work which may include renovations to the original building or additional buildings adjacent to the original that may involve some changes within the original. The originals may have to be retrieved after years of storage, and their condition at that time determines their value. The medium must survive repeated handling without tearing and should not deteriorate.

Intermediates serve a useful function in the drafting process. Properly used, they could save many valuable hours of drafting time.

Always use the correct tools for the specific drafting medium. This includes pens, ink, lead and erasers.

Drawings should always be properly protected. Some offices require that drafting board covers be used to cover drawings during idle periods, days,

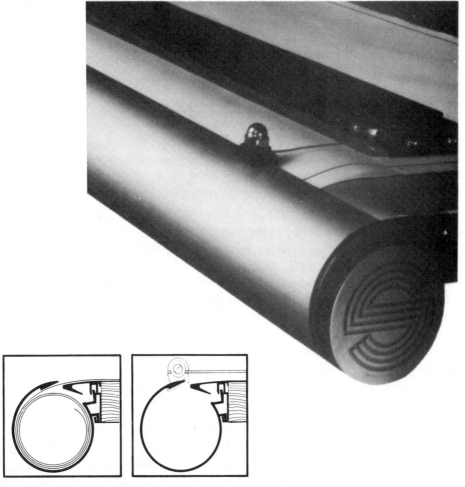

Figure 3-11. SPIROL Drawing Protector mounts on the lower end of drafting tables. By sliding the drawing down into the Spirol, the bottom section is coiled safely out of the way and keeps drawings free from elbow smudges, wrinkling, and torn edges.

nights, and weekends. Other offices require that drawings be removed from the drafting table and stored during idle periods.

Frequently drawings are so place on the drafting table so that the work area is within easy reach of the drafter. In this case, a portion of the drawing may overhang the bottom of the table and be subject to damage by the drafter leaning over the board. Some drafters place a sheet of sketch paper over the drawing at the overhang to protect it. Others use a protector attached to the drafting table. This enables the drafter to work on a drawing while seated or standing in the most natural position. These protectors are made of noncorrosive, extruded aluminum with a clear satin anodized finish. To use it, slide the drawing down into the protector. The bottom portion of the sheet is coiled safely out of the way. It keeps drawings free from elbow smudges and eliminates the need for stretching across the table or working in the shadow cast by the drafter (Fig. 3-11).

Cleanliness is absolutely essential during drafting. Always keep drawings clean and place them on a clean surface. Keep hands, tools, and drafting brush clean. Brush away eraser crumbs and lead point breaks immediately. Avoid unnecessary hand location on the drawing surface, particularly when the hands are moist with perspiration. Cover drawings when not in use. Store drawings in flat files when completed.

PART II
ARCHITECTURAL DRAFTING AND DRAWING DEVELOPMENT

Chapter 4
Introduction

Architectural work in building design starts with client contact. The client may initiate the contact or it may be done by the architect. The client has an idea of what he wants and the architect inputs his vision of the building, and the conception is the result. This part of the work may or may not require any drawings. This is followed by a programming phase which leads directly to drawing development (preliminary phase).

Depending upon the type of project, new or a renovation, some field work, site visits, survey, etc., must be done. This usually includes freehand sketches as the result of field measurements. Drafters are frequently required to perform this work. Programming is usually the architect's responsibility and during this phase he is in contact with the client to ensure that the concept is satisfied. At all phases of design, estimates are developed to verify that the project is within budget. The architect's responsibility may end with the completion of the contract documents or may extend through the construction phase, depending upon the type of contract with the client (Fig. 4-1).

The stages of drawing development where drafters and designers are involved usually consists of three phases—schematics, preliminaries (design development), and final (working) drawings. The degree of involvement by drafters depends upon the level of experience and the practice of the office. Some architects and engineers use entry-level drafters to hard line drawings after the design is sufficiently developed to permit copying, while others expect entry-level drafters to be able to work from rough sketches. In the latter case, drafters are given drawing sheets prepared by architects showing the location of the building and indicating spaces reserved for such essentials as notes, details, key plans, etc.

While the pre-design phase usually does not involve drafters and designers, they should be aware of what is required and how the project is developed before the schematics phase begins. The reason for including such information is that schematics usually begin after there is a meeting-of-the-minds between the architect and client, or owner. This phase is known as programming and includes determining the amount and type of space required, space planning, (juxtapositioning of space), size of proposed building, orientation as necessary, etc.

After the amount and type of space has been determined, the architect develops this program into rough sketches that may include "bubbles" for the space placement with or without a building shape. The next step is the development of a building outline, again a free hand sketch, usually on 8½ by 11 inch sheets or a portion of a 50-yard roll of sketch paper with pencil, ink, or felt mark-

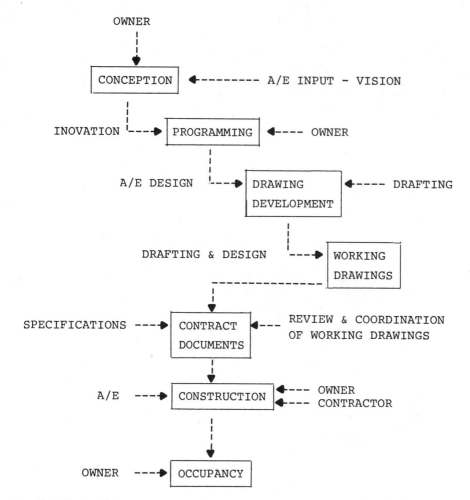

Figure 4-1. Flow diagram illustrates the various phases involved from the initial meeting between the architect and the owner (client) on through occupancy.

ers. How these sketches are developed and what is shown depends upon whether the project is new construction on a clear site, a renovation, or an addition to an existing building. In many cases, these sketches include the site so that the placement of the work can be established and existing conditions can be shown.

Entry-level drafters should be able to take a site plan and develop a hard-line drawing using light strokes for review by the architect. In some instances, drafters and designers, depending upon their experience, may be expected to go to the site alone or with the architect to develop the site plan.

While there is a significant amount of sketching and drawing work performed during pre-schematics, and schematics, drafters really become more completely involved during the preliminary phase of the project which carries through to the final, working drawing phase.

There are many standards (legends and symbols) for use in developing construction drawings that are valuable aids for uniformity and consistency. Most professional offices either have their own or use some national standard. The one weak link in the entire process of developing working drawings is the lack of a standard on good drafting practices. The greatest asset of any entry level or experienced drafter and designer is the use of good drafting practice. Clean, sharp line work and neat, legible lettering are the signatures of good drafters

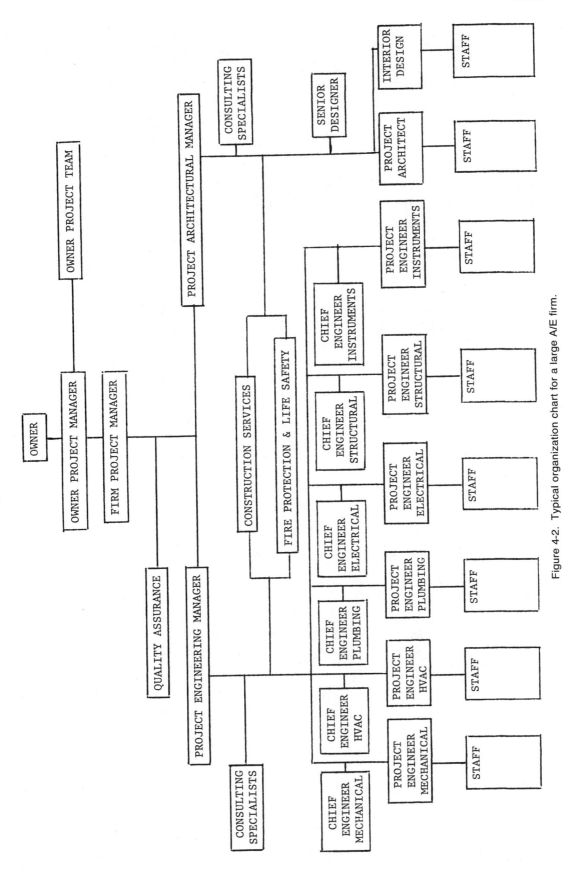

Figure 4-2. Typical organization chart for a large A/E firm.

and designers. These skills are essential for recognition, compensation, and advancement. In addition to these basic skills, all persons working on building drawings must know what is involved in making quality drawings. First, quality drawings must accurately represent that which is to be constructed and second, they must be easy to read, i. e., uncluttered to avoid confusion, accurately dimensioned to eliminate errors in construction, thoroughly coordinated with other discipline drawings and specifications, and all details and section cuts properly identified and easily located.

As entry-level drafters learn and gain experience, they are usually given more latitude in their work and permitted greater freedom of expression. This is not intended to imply building design and redesign. With experience comes less direction and supervision, and drafters are expected to know how to develop quickly and accurately scaled buildings, layout a drawing sheet, understand the need for reserved spaces on drawings, develop good dimensioning practices, and the use and placement of notes.

The architectural drafter must be aware that the building will have a structure and contain electrical and mechanical systems and equipment. The engineering drafter must similarly recognize the fact that the building design is affected by the other engineering disciplines and architecture. All drafters must know that construction drawings are not stand-alone documents (except for very small projects for which specifications are included on the drawings). Drawings and

ARCHITECTURAL AND ENGINEERING DESIGN DISCIPLINES

ARCHITECTURAL	ENGINEERING
ARCHITECTS	CIVIL
INTERIOR DESIGNERS	STRUCTURAL
LANDSCAPE ARCHITECTS	MECHANICAL
ELEVATOR CONSULTANTS	HVAC & AUTOMATION
ARCHITECTURAL ACOUSTICS	PLUMBING
FOOD SERVICE CONSULTANTS	FIRE PROTECTION
	ELECTRICAL
	INSTRUMENTATION
	LIFE SAFETY
	SOILS CONSULTANTS
	TESTING SERVICE
	LIGHTING CONSULTANTS
	LIGHTNING CONSULTANTS
	MECHANICAL ACOUSTICS
	TESTING & BALANCING

Figure 4-3. Architectural and engineering disciplines that may be involved in a building design project.

specifications are complementary documents and must be developed as such. Good drafting practice requires the ability to develop quality drawings that contain all appropriate information and that are fully coordinated with all other design disciplines and the specifications.

Figure 4-2 shows a typical organization chart of a large A/E firm. It is included to show the relationship between the architect and client, and the architect and engineer. The boxes noted as staff include drafters and designers and are usually subdivided to show drafting leaders (supervisors) who may have several drafters, and lead (senior) designers who similarly may have several designers. The structure of the staff depends upon the specific firm.

Figure 4-3 shows individual architectural and engineering design disciplines that may be required on some projects. The various disciplines may be part of the A/E firm or, as in many cases, may be separately retained by the lead professional to perform work in their individual specialty.

There is also the small office in which an architect may be the principle of the firm and will contract for all the additional services, which may also be a small engineering firm. The relationship between the architect or engineer, and drafters and designers is more direct and more often on a personal basis, in small firms. Regardless of the size of the firm, drafters must possess certain basic skills in standard drafting practice and drawing development.

Chapter 5
Types of Drawings

Numerous types of drawings are required from the time the architect or engineer has first contact with the client to when the construction project is completed. These drawings include everything from freehand to hard line work, from sketches on 8½ inch sheets to ink on polyester film. Basic drafting skills are utilized in every facet of the development of building construction drawings.

Various types of drafting techniques are required to illustrate the concept of the finished structure during the different phases from programming through contract documents which include working drawings and specifications. Addendum drawings may be required during the bidding phase and bulletin (modifications) drawings during the construction. Most construction projects consist of five phases, six if programming is included. For the purposes of this discussion the six phases are listed below:

a. Programming
b. Schematic design phase
c. Preliminary (design development) phase
d. Working drawing phase
e. Bidding or negotiation phase
f. Construction phase

1. Programming

Architects are required to provide at least sketches of the building, in some cases, a rendering, and on larger or more complex projects, a scaled model may be required. Drafters and designers are frequently not involved in this phase.

2. Renderings

Renderings are three-dimensional pictorial drawings used to illustrate the project. Frequently it is necessary for architects to prepare pictorial drawings that can be easily understood by persons without technical training. The primary purpose of such drawings is to show buildings as they will appear to an observer after the buildings are completed. Some A/E firms even construct scale models of the project to more clearly define the intended end result of the construction. Modelmaking is a unique specialty and is beyond the planned scope of this book and therefore will be discussed here.

Architectural construction drawings include scaled plans, elevations, sections, and appropriate details sufficient to adequately define the structure so that it can be constructed by a contractor. Pictorial drawings are developed by professionals for use by specialists trained in the interpretation of the drawings. Unfortunately, these drawings are frequently inadequate for many people who lack the specialized training and yet are involved in the various aspects of the building project. Pictorial drawings, also called renderings, are developed to present a picture-like view of the project, be it a single or multi-building complex. Renderings are sometimes a sales tool for the project, but are also used by the A/E firm during drawing development. The perspective drawing is most frequently used by architects, but some prefer the axonometric projection.

There are three types of perspective drawings and these are classified according to the number of vanishing points required. The one-point perspective is most frequently used to illustrate the interior of a space where additional information on the placement of items within the space (room) is considered necessary. A two-point perspective drawing has two vanishing points, one on the left of the drawing and one on the right, and all vertical lines are parallel. This type of drawing is most generally used in architectural drafting to present exterior views of buildings. In some instances, a modification of this type of drawing is used in order to better illustrate special features. A three-point perspective drawing has three vanishing points, two as above, one on the left and one on the right, plus usually one below the drawing. This type of drawing is more difficult to construct and does not warrant the effort since its appearance does not enhance the illustration. Architectural renderings are most frequently constructed using the two-point perspective or a modification of it (Fig. 5-1).

Some architects use what is known as *axonometric projections.* The building is drawn from such a view that three surfaces are shown; this includes the roof and two adjacent walls. First, there are no vanishing points to contend with, and second, all parallel lines are of the same scale. Since the building can be drawn in any number of positions with the walls inclined to the plane of projection, there can be any number of axonometric projections of the same building. Also, since the walls are inclined to the plane of projection, the height of the walls is not shown in their true lengths. The degree of foreshortening of lines depends on their inclination to the plane of projection (Fig. 5-2).

There are excellent texts available for those drafters who desire to explore further the subject of rendering drawings (perspectives and axonometrics).

3. Schematics

The schematic design phase is the first phase of building construction drawing development and is also the first stage of involvement for drafters. A schematic drawing is basically an outline drawing showing the bare essentials of the structure. Some firms do all schematic work on 8½ × 11 inch sheets, while others use full-sized drawings. Schematics illustrate the scheme of the structure and are handled differently by different firms. The drawings developed at this stage are never used by contractors for building construction, but the quality of drafting should satisfy the intended purpose, that is, the line work should be sharp and clear, and the lettering should be legible.

The type of drawings developed depends upon the scope of the project and practice of the firm. A small architectural office (one man shop) may start by making rough sketches on any type of paper available during the initial contact with the client; larger firms are usually more formal in the initial sketch development. In either case a basis of design will be developed for presentation to

Figure 5-1.   Rendering by Sidney Scott Smith, AIA, Architect. Reprinted by permission.

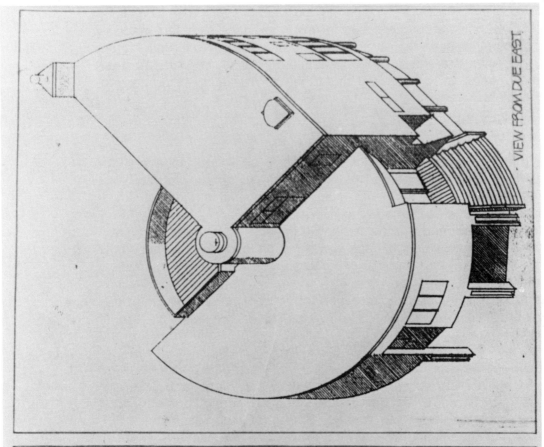

Figure 5-2. Axonometric drawing by Sidney Scott Smith, AIA, Architect. Reprinted by permission.

the client for approval. This phase could include a simple floor plan with elevations on 8½ × 11 inch sheets developed for initial freehand sketches made by the architect. Whether this work is merely basic drawings or a thoroughly detailed basis of design, the end result is to adequately define the intended structure in sufficient detail to convince the client that the architect has properly interpreted the clients needs.

4. Preliminary Drawings

Preliminary drawings are usually on full-sized sheets similar to those used for working drawings. These are primarily single-line drawings showing architectural and engineering design in greater detail. This is actually the design development phase and includes the following information:

a. Architectural—floor and roof plans showing space assignment, sizes and outline of fixed and movable equipment; elevations and typical sections; shafts and stairways; site plans showing roads, parking areas and sidewalks, utilities, and site conditions and constraints.

b. Mechanical—single line layouts of all duct and piping systems; riser diagrams where applicable; scale layout of boilers and major associated equipment and central heating, cooling and ventilating units; fire protection system; and plumbing piping and riser diagrams.

c. Electrical—plans showing space assignment, sizes and outline of fixed equipment such as transformers, main switch and switchboards, and generator sets; riser diagram showing arrangement of feeders, subfeeders, bus work, load centers, and branch circuit panels; and security and fire alarm systems.

d. Structural—basic building structural systems; column locations; footings and foundations.

The information shown on preliminary drawings must be sufficient to describe the project adequately. The design is frequently done at ¹/₁₆ inch scale, but then these drawings are usually *discarded* after the preliminaries are completed and approved by the client. The drawings are *discarded* to reference use only and new ⅛ or ¼ inch scale drawings are made for working drawings. On smaller and less complex projects, the preliminaries are done at the scale intended for working drawings; for example, ⅛ or ¼ inch scale. These drawings are then developed to the point necessary to describe the project adequately, and obtain approval from the client. After such approval, work is continued on these drawings until they are fully documented and detailed for use as working drawings. In this way no drawings are wasted and usually design time is reduced and money is saved for the A/E.

5. Working Drawings

Working drawings are actually the end result of the entire drafting and design effort. These are the drawings used by contractors to construct the structure. From 40 to 70% of the design time and total fee is consumed during the working drawing phase, therefore it deserves equal treatment in this book. The techniques of drafting practice and drawing development explained here can be used for all other phases.

After the preliminary (design development) drawings have been completed and approved by the client, they are used as the basis for the development of working drawings. A recent contract between a client and architect best describes what is expected of working drawings *WORKING DRAWINGS SHALL BE COMPLETE AND ADEQUATE FOR BIDDING, CONTRACTING, AND CONSTRUCTION PURPOSES.*

While working drawings include all architectural and engineering design, this chapter addresses architectural drawings only; engineering drawings are covered in Part III.

a. Sheet Sizes

Drawing sheets are available in a wide variety of sizes ranging from the 8½ × 11 inches to 50-yard rolls. A recent survey of manufacturers of tracing paper, vellum, and polyester film showed that the following sizes are available:

$$8\frac{1}{2} \times 11$$

$$9 \times 12$$

$$11 \times 17$$

$$12 \times 18$$

$$18 \times 14$$

$$22 \times 34$$

$$24 \times 36$$

$$30 \times 42$$

$$36 \times 48$$

and in roll form of 30, 36 and 42 inch widths in lengths of 20 and 50 yards. The cut sheet sizes are available in plain sheet form or preprinted with borders and title blocks.

The 8½ × 11 inch and the 11 × 17 inch sizes are most frequently used whenever these are included in bound copy or for schematics, addenda and bulletins. This size can also be conveniently stored in file folders. The 24 × 36 inch and 30 × 42 inch sizes are most frequently used for working drawings. Either of these two sizes, in the plain sheet form, can be cut in half to produce 18 × 24 inch and 21 × 30 inch sheets which yield sufficient variation for most drawing needs, with the exception of very large drawings that are difficult to handle by the contractors on the job.

The sheet size selected should be large enough to permit the placement of the building on one sheet. In cases where the building is too large for one sheet, the building may be placed on two sheets, with the building cut into approximately two equal sections. Never use a sheet size where a small portion, say less than 30% is placed on a second sheet. There is a standard technique for indicating the cut section of a building. This is described later. Also, never use a sheet size where the building completely fills the sheet; always leave room for appropriate dimensioning and notes that must be placed outside the building exterior lines. Working drawings have a border which is set by office standard practice, usually a heavy line ½ or ¾ inch in from the outer edge, all around the sheet with an additional 1½ inch border for binding on the left margin. Some offices use preprinted sheets complete with border and title for uniformity of work, in two or three standard sizes.

b. Sheet Layout

After a building scale has been determined, it is necessary to properly place the building on the drawing.

Figure 5-3 shows a typical building placed on a standard 24 × 36 inch sheet. This building is placed approximately in the center of the sheet after space has been set aside for binding on the left, notes on the right, and reserved space and title block on the bottom. This plan leaves adequate space for dimensioning and some essential details that may be required on the same sheet.

c. Scaling Buildings

As stated earlier, working drawings are generally done at ⅛ or ¼ inch scale depending upon many different situations and circumstances. To be able to quickly and accurately determine lengths in inches that represent scaled reductions of buildings sized in feet is a definite requirement for all drafters and designers. Take for example, a building 160 feet long by 112 feet wide and the assignment to layout the building on a 24 × 36 inch sheet leaving appropriate reserved spaces as shown in Figure 5-3. What scale best satisfies the problem? The free space of the drawing would be about 20 × 27 inches. What scale is best; 1/16, ⅛, or ¼ inch?

The scale of a drawing converts building dimensions from feet to inches in

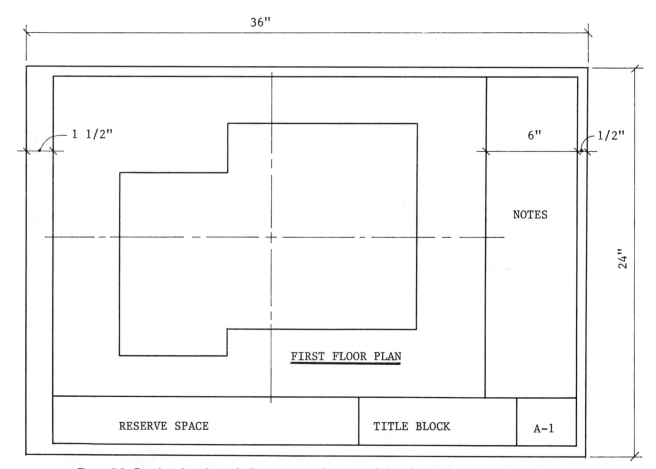

Figure 5-3. Drawing sheet layout indicates proper placement of plan views with appropriate space retained for notes on the right edge space on the left for binding the drawings.

accordance with the scale used. For example, with a scale of ¼ inch = 1 foot-0 inch (¼" = 1'-0"), each inch on the drawing represents 4 feet of actual building dimension. From this it is simple enough to convert from building dimensions to drawing scaled dimensions, or at least so it seems. In case there are any questions as to how this is done, the following should eliminate any confusion.

Back to the 160-foot long building, at ¼ inch scale, how many inches are needed on the drawing to draw this length? The obvious answer is 40 inches and the conversion is rather easy. For example, multiply the actual building dimension in feet by the inch fraction of the scale used.

$$@ \ \tfrac{1}{16}—160 \times \tfrac{1}{16} = 10 \text{ inches}$$

$$@ \ \tfrac{1}{8}—160 \times \tfrac{1}{8} = 20 \text{ inches}$$

$$@ \ \tfrac{1}{4}—160 \times \tfrac{1}{4} = 40 \text{ inches}$$

Similarly, for the other dimension:

$$@ \ \tfrac{1}{16}—112 \times \tfrac{1}{16} = 7 \text{ inches}$$

$$@ \ \tfrac{1}{8}—112 \times \tfrac{1}{8} = 14 \text{ inches}$$

$$@ \ \tfrac{1}{4}—112 \times \tfrac{1}{4} = 28 \text{ inches}$$

In answer to the question relative to the building scale, the ⅛ inch scale best suits sheet size selected for the drawing.

To take a drawing scaled dimension and convert it into actual building dimensions, reverse the process used above. Divide the scaled dimension by the scale fraction: 14"/⅛ = 112'. A simpler method is to use the denominator of the fraction, i. e., divide/multiple the dimension by the denominator of the scale:

$$\text{to reduce feet to inches } 112'/8 = 14".$$
$$\text{to enlarge inches to feet } 14" \times 8 = 112'.$$

d. Using Scales

In order to illustrate a building on a handy sized sheet, the building must be *scaled* down to the appropriate size to fit on the sheet properly. For this reduction to be of value, the building must be drawn in exact proportions, i. e., when a building is drawn at a scale of ¼" = 1'-0", the result is a drawing of the building $\tfrac{1}{48}$ the actual dimension. This means the same reduction for every portion of the building. Various scales are used in the process of making a set of drawings, for example, the floor plan may be at ¼ scale, and details may be drawn at ½ scale, or even larger where necessary. Drafters must become adept in the use of scales.

There are two types of scales used in building construction drafting, the architect's scale and the engineer's scale. The architect's scales are used for all work other than site plans and other similar drawings. The difference between the two is that the architect's scale is designed to measure dimensions in feet and fractions of feet and inches, while the engineer's scales are designed to measure in feet and decimals of feet. On the architect's scale 1" = 1'-0", there are 48 divisions between the inch mark, representing ¼₈th of a foot per mark, while on the engineer's scale, 1" = 1', there are 10 divisions between each inch mark, representing 0.1th of a foot per division.

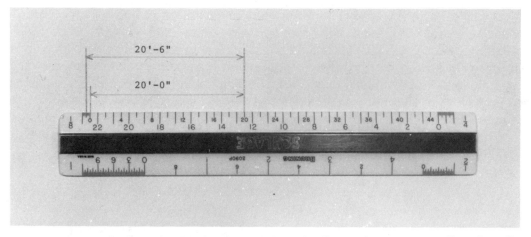

Figure 5-4. Architect's, 6-inch, four-bevel pocket scale is used to illustrate how to measure dimensions. On the scale: ⅛″ = 1′0″, each division to the right of 0 represents one foot. Each division to the left of 0 (there are six) represents 2 inches.

Figure 5-4 shows how to measure scaled dimensions with an architect's scale where ⅛″ = 1′-0″. To measure 20′-0″ (read twenty feet and zero inches) measure from the 0 mark to the 20 mark. This method applies to any dimension where the length is in full feet. To measure 20′-6″ (read twenty feet and six inches) measure from the mark representing 6 inches to the left of 0 on the 20 mark. The ⅛ scale has six divisions to the left of 0 with each division representing 2 inches. The ⅛ scale reads from the left to the right. In order to provide a more versatile tool, two scales are marked on each face. On the same face as the ⅛th scale the ¼th scale is also indicated. This scale reads from right to left and the dimensions are indicate by the upper set of numbers. To the right of 0 on the ¼ scale there are 12 divisions, and as the scales become larger, the subdivisions increase, providing greater accuracy when making larger scaled drawings.

Conversely, to determine the length of a line on a scaled drawing, select the appropriate scale and measure the line. For example, what is the length of the line in feet and inches between the two outer lines on Figure 5-4 at the scale of ⅛″ = 1′-0″? Since the length of that line is such that it does not fall on exact feet divisions, set the scale so that the 20-foot mark matches the right end of the line and read the fraction in inches to the left of 0. This method must be used whenever reading dimensions that are not in exact foot dimensions.

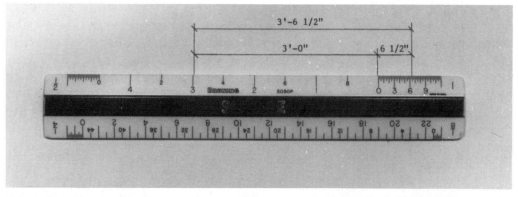

Figure 5-5. Using the scale: 1″ = 1′0″, each division to the left of 0 represents 1 foot. There are 48 divisions to the right of 0 with each representing ¼ inch, and the numbers to the right, 3, 6, and 9, represent inches.

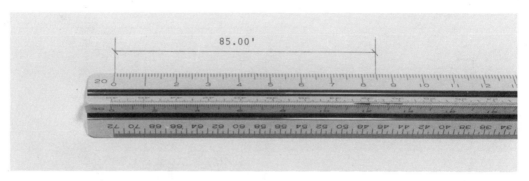

Figure 5-6. Engineer's (decimal) scale is used for site plans and other similar drawings. On the scale: 1" = 20', each division represents one foot with the Number 1 representing 10 feet, No. 2 = 20 feet, etc. The measured distance is 85.00'.

Another difference between the two types of scales is that the engineer's scale starts at 0 and reads in one direction only. For example, using the scale 1" = 20' as shown on Figure 5-6, set 0 on the left line and read to the 5th division past the 8 mark (each mark has 10 divisions) therefore this point represents 85.00 feet.

e. Dimensioning Methods

A thorough knowledge of the fundamentals of good dimensioning practice is essential for all drafters and designers. It is just as easy to learn to dimension correctly at the outset and thereafter use the proper method. While building construction drawings are made to scale, do not expect contractors to scale the drawings, and be aware that under various atmospheric conditions, print paper is likely to shrink. The drawing is merely a picture showing sizes and relationships, and figure dimensions establish the exact size. In the event of error or confusion, figure dimensions take legal precedence over scaled dimensions. Good dimensioning practice is essential to reduce the probability of error.

All drawings that are drawn to scale must show the scale used in the appropriate location, adjacent to the portion scaled. Overall dimensions are usually placed below and to the right of the drawing. Some dimensions must be placed within the drawing. The most important consideration is that the dimensioning lines be readily identifiable and not confused with the building or equipment lines. Some details are not drawn to scale and are usually so identified by the notation NTS (not to scale). Isometric piping drawings, plumbing riser diagrams, and electrical single-line diagrams are not drawn to scale.

There are two types of lines associated with dimensioning—extension lines and dimension lines. Additionally, there are generally three levels of dimensioning used on plan and major drawings. The line nearest to the building identifies *openings* and *corners;* the next identifies *major features;* and the last, furthest from the building shows the *overall dimension.* On sections, details, etc., frequently only one or two lines are required. (Fig. 5-7)

The following are important considerations for good dimensioning practice:

1. Extension lines are placed perpendicular to but not touching building lines; they are light lines as compared to the building and equipment lines.

2. Dimension lines are placed parallel and adjacent to building and equipment lines. Dimension lines terminate at extension lines with an appropriate terminating symbol. (Fig. 5-8). Like extension lines, these are light lines.

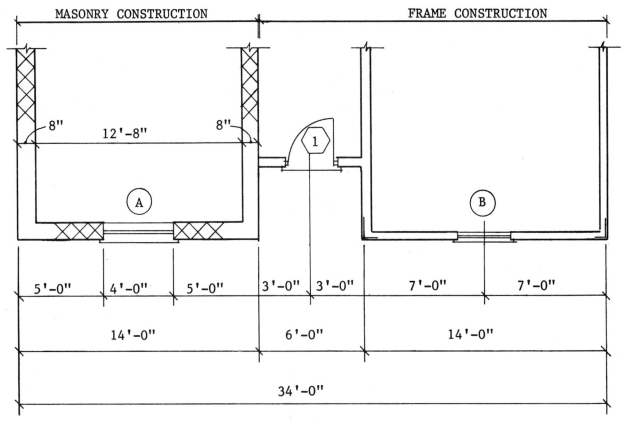

MASONRY CONSTRUCTION

FRAME CONSTRUCTION

8" 12'-8" 8"

A

1

B

5'-0" 4'-0" 5'-0" 3'-0" 3'-0" 7'-0" 7'-0"

14'-0" 6'-0" 14'-0"

34'-0"

Figure 5-7. Good dimensioning practice usually requires three dimension lines for building plan views. The first line nearest to the building is for openings and corners, the second line is for major features, and the outside line is for overall dimensions.

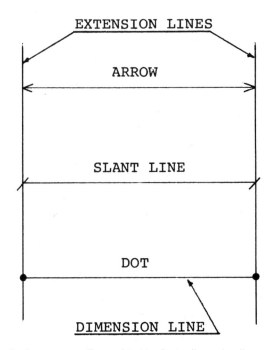

EXTENSION LINES

ARROW

SLANT LINE

DOT

DIMENSION LINE

Figure 5-8. Three methods are generally used to terminate dimension lines; arrow, slash, and dot.

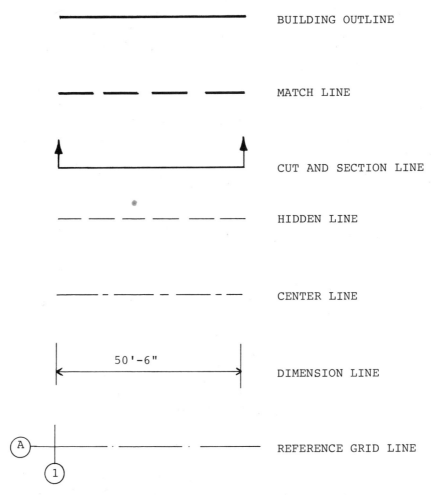

BUILDING OUTLINE

MATCH LINE

CUT AND SECTION LINE

HIDDEN LINE

CENTER LINE

50'-6"   DIMENSION LINE

A   REFERENCE GRID LINE
1

Figure 5-9. Various types and weight (width) of lines are necessary to properly identify different segments of drawings. Varying the types of lines enhances the quality of the drawing.

3. All dimensions are placed above the dimension lines and are written in feet and inches, as 1'-4", etc., except when the dimension is 12 inches, then it may be indicated as 12" or 1'-0".

4. When there is insufficient space for the dimension, then the preferred method is to use a leader from the space to the dimension which is placed outside of the space.

5. When dimensioning a building work along the full length of the scale instead of moving the scale for every different division to eliminate the possibility of cumulative errors.

6. Fractions are usually written with a diagonal slash line—¼, ½, and ⅛. This method is usually easier to print and takes less space vertically.

7. All dimension numerals and fractions should be large enough to read easily and must be legible to avoid confusion and errors.

8. Architectural scales are used for dimensioning buildings and equipment; engineering scales are used for dimensioning site plans, etc.

f. Type and Quality of Lines

Drawings consist of lines and lettering, and the combination of these two components, properly executed, convey the design intent correctly, clearly, and sim-

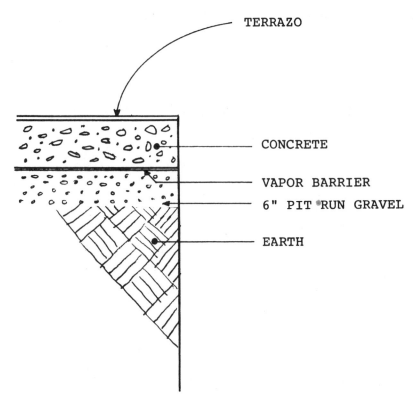

Figure 5-10. Leader lines are used to identify materials of construction, application points of notes, and identification of equipment. Arrows must terminate (touch) the material illustrated or point of application. Dots are used to terminate in the material.

ply. Just as several different sizes of letters and numerals are used, where appropriate, different types and quality (weight) of lines are necessary to illustrate their importance graphically. Since the intent of drawings is to convey information in an intelligible manner, a diligent effort is necessary to make the graphics readily comprehensible.

Various sizes of lettering are necessary to focus attention upon the essentials such as titles and headings. Also, different weight lines emphasize important features of the drawings. Very simply, dark lines indicate important elements, light lines indicate lesser important elements. Also, different types of lines, for example, solid, broken and dashed lines, are used to indicate different aspects, features, and elements of the drawings.

Architectural drawings most frequently depend upon weight and type of line to show outlines and graphic symbols to indicate construction materials.

Figure 5-9 illustrates the various common types and weights of lines most frequently used on architectural drawings. Other types may be used when necessary, as long as they are appropriate. Whenever there are any questions as to the intent of any line used, the line intent should be identified.

Figure 5-10 illustrates the method used when indicating the materials of construction outside the boundary of the drawing.

Chapter 6
Architectural Working Drawings

For the purposes of this book, the architect is considered the prime professional, i. e., the architect has the contact and contract with the client and the architect subcontracts all engineering and other design work. The architect progresses through all preliminary planning and drawing development stages, in coordination with the engineers, to ensure that the planned structure can be constructed and that appropriate spaces are provided for all necessary engineering equipment. After completion of the preliminary (development) drawing phase, and approval by the client, design begins on working drawings.

Drawings are generally identified by a letter prefix in the title block, for example. "A" for architectural drawings, "E" for electrical, etc., but this does not mean that all the information on the specific drawing is the product of that discipline. A good example is the architectural reflected ceiling plan which is issued as part of an architectural set but will contain input from electrical, mechanical and fire protection, as required.

Architectural drawings here will be considered as those that are identified by the letter prefix "A." The site plan is shown as part of the architectural set but on some projects there may be as many as three additional site plans developed by the engineers and included in their own sets.

This section will deal with architectural drawings as listed below:

1. Title sheet
2. Site plan
3. Floor plan
4. Roof plan
5. Reflected ceiling plan
6. Elevations
7. Sections
8. Details
9. Schedules

All drawings must contain certain basic, essential information:

1. Name and address of the project
2. Name and address of the architect or engineer
3. Sheet title
4. Sheet number
5. Project number
6. Date project is completed

7. Space for initials of person making the drawing
8. Space for initials of person checking the drawing
9. Mark, date and description of all revisions
10. North arrow; this may include true north and building north
11. Key plan (Fig. 6-3)
12. Vicinity map (Fig. 6-4)

Title blocks are required on all drawings (except renderings) and similar information must appear on the title page of all specifications prepared and sealed by architects and/or engineers. The design of the blocks and the information required may vary somewhat from one state to another, but certain basic information either *should* or *shall* be included.

Title blocks must be distinct and separate from any other title block, box, plaque, or other similar device of illustration or lettering. Drawing mediums may be purchased with a *standard* title block, or *custom* title blocks may be ordered with the drawing medium or separately with adhesive backing to be placed on plain drawing sheets. Title blocks are placed along the bottom of the sheet, bottom right-hand corner, or along the right-hand edge reading from the bottom up (Fig. 6-1).

Space must be provided either in the title block or adjacent to it for *sealing* by the architect and/or engineer. Some states require that all original drawings be sealed with an inked stamp; others require that prints be sealed by a signature and indent stamp, and still others require that the professional sign it and write his license number on the drawings or prints. These requirements vary from state to state and may change with time.

Title blocks are an essential part of working drawings, and uniformity of the title blocks and the manner in which they are completed enhance the professional quality of a set of drawings. All lettering on the set should be done by a similar method, i. e., pencil or ink, freehand, template, press-type lettering, or mechanical lettering device. Even the seemingly insignificant uniformity of the date of the drawing is important, for example, when the architect decides that Dec. 31, 1984 is to be used, drafters should use that format on all drawings of the set and not Dec. 31/84, or 12/31/84, etc., on different drawings of the set.

1. TITLE SHEET

Most projects contain a title sheet as the first sheet of the drawing set. What is included on the title sheet as the first sheet depends upon the type and size of the project. For example, on smaller projects, the sheet may contain only the name and location of the project in large bold letters up to 2 inches high for the title or name of the project and letters up to 1 inch for the address. On larger and more involved projects, a site plan may be shown along with the symbols used for the site plan, sheet index, vicinity map, and key plan. The key plan is frequently shown on all drawings of the set (Fig. 6-2).

2. SITE PLAN

Site plans usually contain all appropriate site information detail including building outline, lot (property) lines, existing and new grade contour lines, power and water lines, trees to be removed and remaining, storm drainage, catch basins and manholes, building floor slab elevations, paved areas, sod and/or seeded areas. Space permitting, this drawing should also contain all appropriate legends pertaining to the site plane (Fig. 6-5).

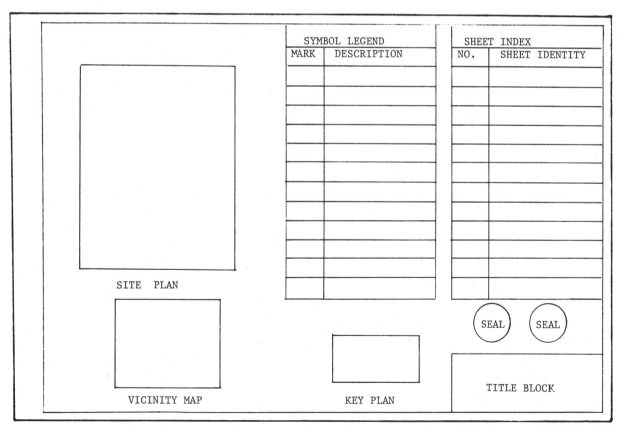

MARK	DATE	DESCRIPTION OF REVISION

ARCHITECT'S OR ENGINEER'S NAME

SCALE:	APPROVED BY:	DRAWN BY
DATE:		REVISED

WORK OR PROJECT TITLE AND ADDRESS

DRAWING TITLE	DRAWING NUMBER

Figure 6-1. Drawing sheets may be purchased with preprinted title blocks. Space for revisions should be added, if not included on preprinted format, or at least reserved, before any drafting work is started.

SITE PLAN

SYMBOL LEGEND

MARK	DESCRIPTION

SHEET INDEX

NO.	SHEET IDENTITY

VICINITY MAP

KEY PLAN

SEAL SEAL

TITLE BLOCK

Figure 6-2. The title sheet is usually the first sheet of the set, and as such should clearly identify and locate the project. A title sheet could contain: site plan, site plan symbol list, vicinity map, key plan, and a sheet index listing all the drawings of the set.

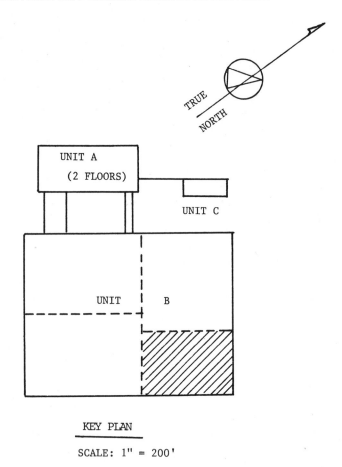

KEY PLAN

SCALE: 1" = 200'

Figure 6-3. Key plans are small scale-drawings of the building and are usually shown on all drawings. Cross-hatching may be used to indicate that portion of the building involved in the project when less than the entire building is included.

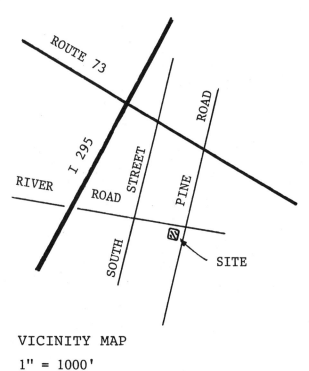

VICINITY MAP

1" = 1000'

Figure 6-4. Vicinity maps are small-scale maps of the area used to locate the site of the project when direction to the site is considered necessary.

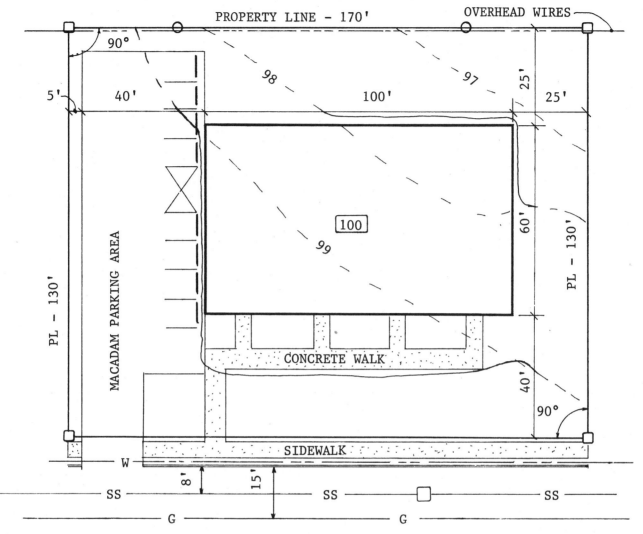

Figure 6-5. Site plans are used to locate the building on the property, define site constraints, locate site services, and establish property and building lines.

Site plans are frequently one of the first drawings completed, since they define site constraints, locate all site services, and establish property lines. On small projects the architect usually includes the site plan in the "A" set, while on large projects the various engineering disciplines produce individual site drawings. Sometimes site plans are combined to show all architectural and engineering information and then these drawings are so identified, as "AEP-1", indicating that architectural, electrical, and plumbing information is included.

3. FLOOR PLANS

Drawings are generally divided into two groups—horizontal (plan) views, and vertical (elevation and section) views. Plan views are usually titled by what is shown as floor plans, reflected ceiling plans, and roof plans instead of floor plan view, etc., the word view is implied. These views are drawn to scale and the various building materials, equipment, and fixtures are indicated by symbols or notes, whichever is more appropriate for the situation. Sometimes a combination of symbols and notes are used to reduce actual drafting time.

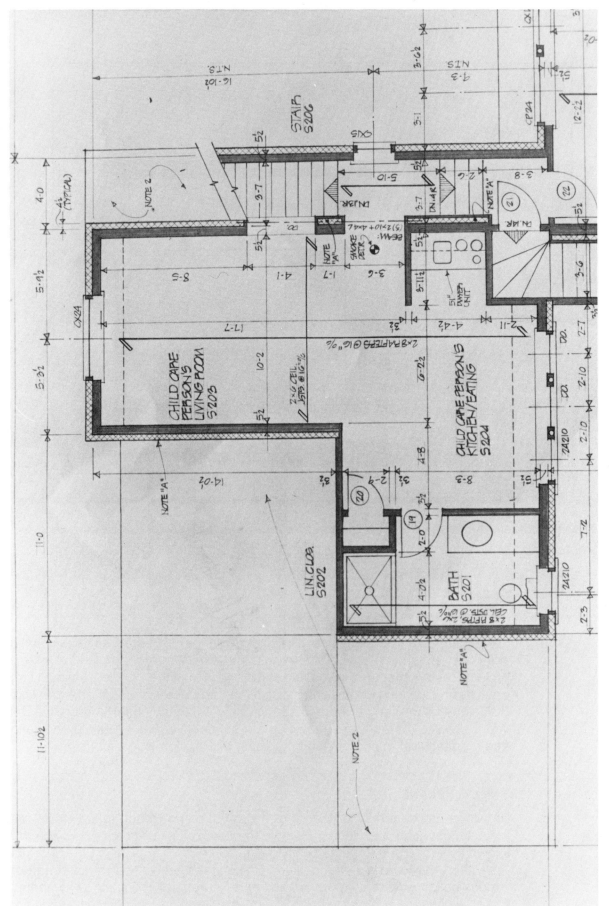

Figure 6-6. Partial residential building floor plan. Note dimensioning technique and proper use of symbols. Notes A and 2 refer to descriptions on list of notes included on drawings. (Courtesy of Sidney Scott Smith, AIA, Architect. Reprinted by permission.)

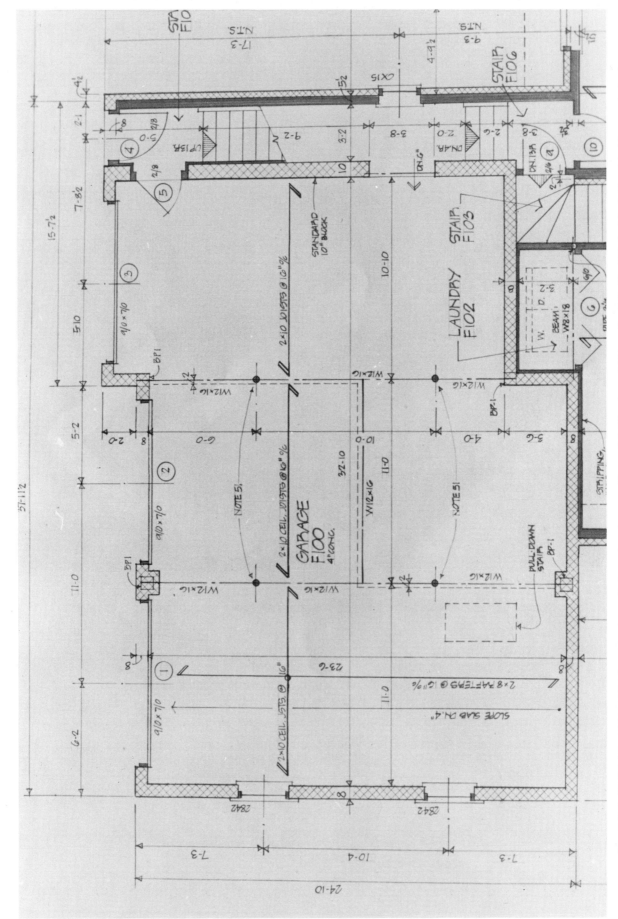

Figure 6-7. Partial ground floor plan for space beneath floor plan shown on Figure 6-6. (Courtesy of Sidney Scott Smith, AIA, Architect. Reprinted by permission.)

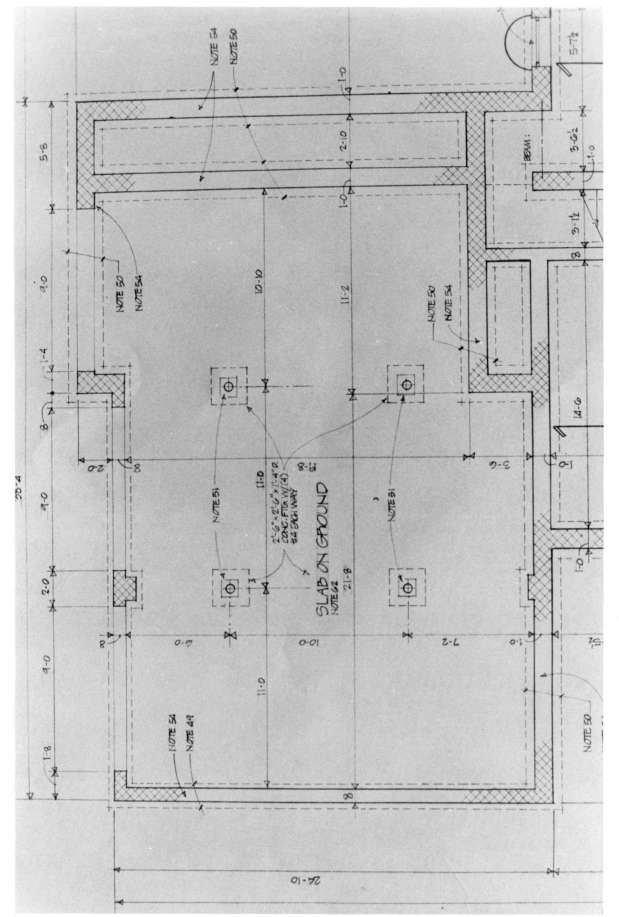

Figure 6-8. Partial foundation plan is uncluttered through the judicious use of notes. (Courtesy of Sidney Scott Smith, AIA, Architect. Reprinted by permission.)

When the scale of a building is large enough to show walls by two lines, symbols are used to indicate the various construction materials. Remember, specifications describe the type of dry wall, block, concrete or brick as intended, and the drawing should merely indicate the different types (Fig. 6-11).

Symbols are similarly used for windows and doors. These are usually identified by a *number* or *letter* appropriately enclosed and the type is described in the schedule. (Fig. 6-9, Fig. 6-10).

The floor plan is of primary importance for the development of working drawings. It is the drawing from which all other architectural and engineering design is done. It is used to develop exterior and interior elevations, sections, and appropriate details. Working drawing floor plans are the result of all prior design effort and the culmination of many meetings with the client and in coordination with the engineers.

A floor plan is a horizontal view of the building taken at an appropriate level

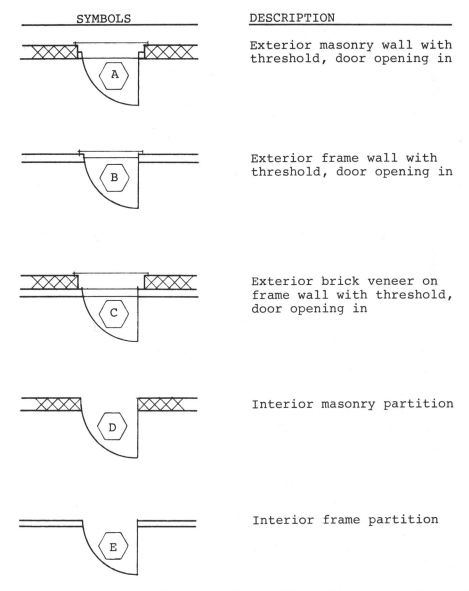

SYMBOLS	DESCRIPTION
A	Exterior masonry wall with threshold, door opening in
B	Exterior frame wall with threshold, door opening in
C	Exterior brick veneer on frame wall with threshold, door opening in
D	Interior masonry partition
E	Interior frame partition

Figure 6-9. Typical door symbols used in plan views. The type of door and frame is described in door schedules or specifications.

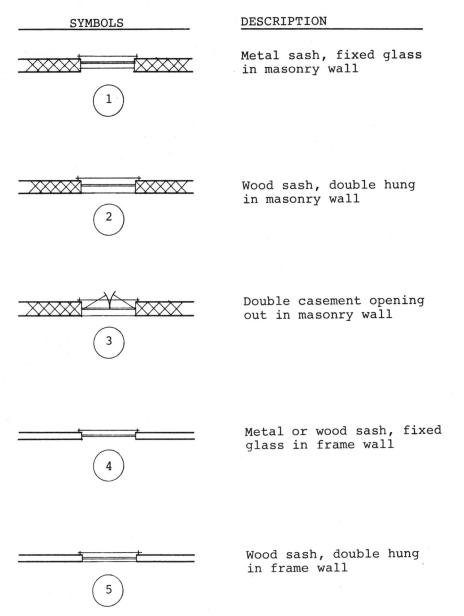

SYMBOLS	DESCRIPTION
1	Metal sash, fixed glass in masonry wall
2	Wood sash, double hung in masonry wall
3	Double casement opening out in masonry wall
4	Metal or wood sash, fixed glass in frame wall
5	Wood sash, double hung in frame wall

Figure 6-10. Typical window symbols used in plan views. The type of window is described in window schedules or specifications.

so that all openings in walls and partitions will be shown. Floor plans are fully dimensioned, showing interior and exterior dimensions, wall thickness, and room and space sizes. On small projects, rooms and spaces are identified by name; on large multi-story buildings, room numbers are used, such as 100, 101, 102 for the first floor; 200, 201, 202 for the second floor, and so on through the building.

The following are some of the items generally included on floor plans:

1. All wall types are identified by appropriate symbols and keyed to the wall symbol list.
2. Wall thickness is dimensioned.
3. Door openings showing door swing and type identifications.
4. Windows with type identifications.

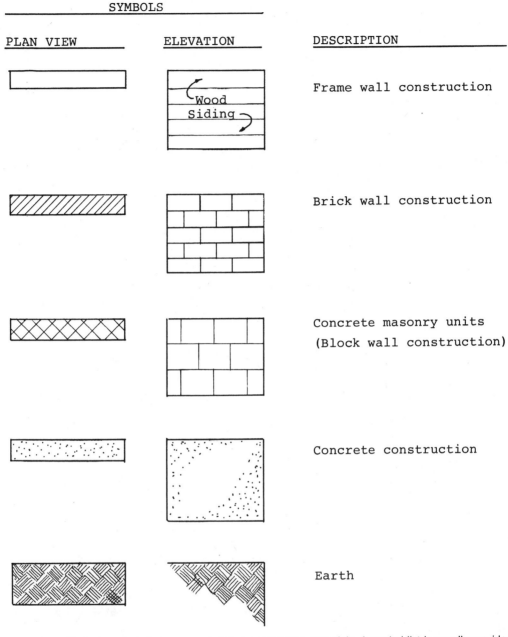

Figure 6-11. Typical symbols for common building construction materials. A symbol list is usually provided to eliminate error. In elevation, wood siding may be partially illustrated with a note indicating type, "siding", "shingles", etc. Partial illustration is frequently used in both plan and elevation views for brick, block, concrete, and earth to save drafting time.

5. Equipment below window openings.

6. Location of all wall section cuts and interior elevations keyed to drawing where they appear.

7. Concrete slabs and steps at exterior doors.

8. Roof overhang may be indicated by light dashed lines.

9. Plumbing fixtures.

10. Floor drains and slopes where necessary.

11. Kitchen equipment (except if extensive, then show on a separate enlarged drawing).

12. Water fountains, built-in and surface cabinets.
13. Wall-mounted items, bulletin boards, and chalk boards.
14. Fire extinguishers by type and locations.
15. Telephone outlets on small projects.
16. Floor elevations.
17. Room identifications.

The amount of information included on floor plans depends entirely on the scope of the project and the complexity of construction. This is the drawing used by contractors to install the floor and to layout and construct walls and partitions. Where additional detail is required, these areas are identified and keyed to the drawing where this information is located. The essential consideration in drawing a floor plan is to know how much to show and that depends on the scale of the drawing and the complexity of construction. A typical set of architectural working drawings contain elevations, sections, and enlarged scale detail drawings to provide additional information related to the floor plans. For example, exterior elevations and wall sections are always used to describe wall construction. Enlarged detail plans are frequently used to show necessary details of toilet rooms, locker rooms, and similar special areas. Sections and details are usually required to describe construction materials and methods at entrances. Elevations and sections are generally required for stairs. Enlarged details are required at various interconnecting points of construction.

On a ⅛th inch scale less can be shown and more detail is needed. On a ¼ inch scale more can be shown and less need be detailed. The drafter and designer need to understand the use of elevations, sections, details, and schedules to develop floor plans properly and adequately. Never draw more than is necessary.

Floor plans should be made at a scale large enough to show all essential information on a drawing that is not over-sized for convenient handling in the drafting room and in the field during construction. When these two conditions cannot be met, the building must be *cut* and the parts placed on two separate sheets. A match line is used to indicate where the building is cut. This line is much darker than the building lines to avoid confusion, and the match line is appropriately identified. Once the decision has been made that the building must be cut, all subsequent and relevant drawings must be so shown. This includes architectural roof and reflected ceiling plans, and all associated engineering drawings.

The architectural drawings that follow the floor plans depend to some extent upon office practice. It could be followed, as indicated here, by the roof plan, reflected ceiling plan, exterior elevations, and wall sections. In larger offices, all of these may be worked on simultaneously. Roof plans and reflected ceiling plans are usually developed concurrently with the appropriate engineering design because their input is required before these two drawings can be completed. Exterior elevations and wall section designs are usually performed at the same time. The wall sections show the construction of the wall, and where exterior brick, block, etc., coursing type material is used, it is necessary to determine the locations of the various courses in order to define the elevation fully. What follows is not intended to illustrate the sequence of drawing development but rather to describe the various types of drawings, what they are, what must be shown, and their relationships to other drawings.

4. ROOF PLAN

While the roof plan is usually one of the simplest drawings to make, certain essential details must be properly illustrated to ensure a weather-tight construction. A waterproof roof is absolutely necessary for the building since it is the ultimate weather shield or umbrella; if the roof fails, so does the building. The result of a poorly designed and constructed roof is more obvious than almost any other element of the building. A roof plan is required for all by the simplest structure, except residences, and similarly sloped roof buildings (Fig. 6-12).

The roof plan is usually drawn at the same scale as the floor plan and is the view as seen from above the roof. While architects make this drawing, it shows all equipment on the roof, including mechanical, electrical, etc., but the details and descriptions are included on those plans made by the various engineering disciplines. The architect shows architectural equipment and frequently describes and dimensions them as required. The roof plan must show all roof penetrations and include the slope of the roof. Roofing materials and methods of installation may be included on the drawing or included in notes or in the specifications. Some drawings indicate in phantom the structure below the roof, such as joists, beams and columns, for information only. Joists and beams are usually partially shown to indicate location and not the complete length of the roof. Roof plans should not be confused with roof framing plans; these are separate drawings and show the structural framing as it appears beneath the roof. On larger and more involved projects, the engineering disciplines frequently make roof plans that address that specific engineering discipline's work.

Architectural roof plans usually show overall dimensions and dimensions to specific architectural features such as hatches (scuttles), expansion joints, penthouses, skylights, roof ventilation, walkways, scuppers, etc. These plans also show all roof penetrations and all equipment on the roof.

Most frequently, equipment and materials that penetrate or are on the roof are indicated by symbols or abbreviations because they are described on other drawings or in schedules, details, or specifications. On small projects, where there is only one roof plan drawn by the architect, the equipment may be fully described on that plan or by notes on the same drawing.

One of the major concerns of architects is the method of terminating the roof at parapets, expansion joints, equipment curbs, and at the various penetrations. These details are usually drawn at an enlarged scale on the same sheet, space permitting, to ensure proper construction methods at these points. The amount and number of details depend upon office practice and the complexity of the work (Fig. 6-13, Fig. 6-14, Fig. 6-15, Fig. 6-16, Fig. 6-17).

These details shown are for specific applications and cannot be used for all installations; they are for illustrations only. The methods used must agree with manufacturer's recommendations.

5. REFLECTED CEILING PLAN

The reflected ceiling plan is a composite drawing made by the architect and is included in the architectural drawing set. It contains, in addition to the ceiling which is an architectural feature, electrical and mechanical items that penetrate the ceiling and are visible from below. The architectural features include ceiling type and material, and methods of attachment to all adjacent surfaces such as walls, windows, columns. Engineering items may include lighting fixtures, HVAC

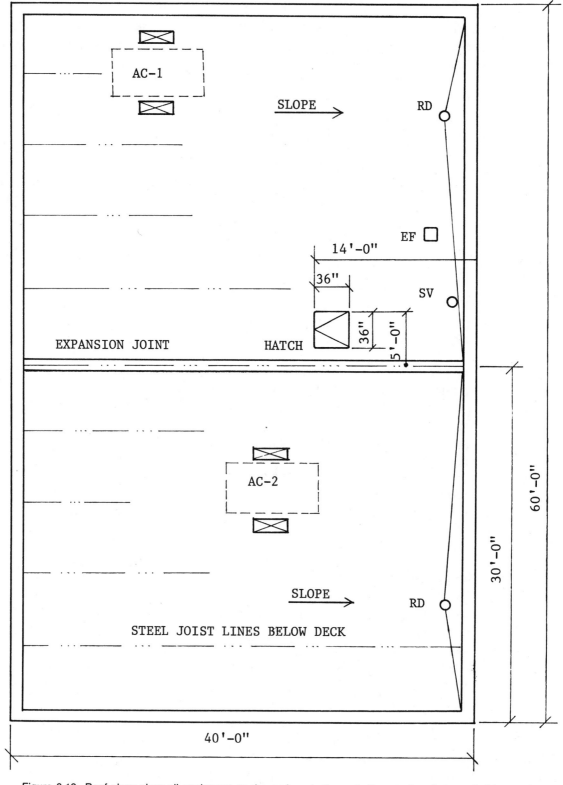

Figure 6-12. Roof plans show all equipment on the roof, and all penetrations and roof slope. Architectural equipment is usually dimensioned. Plumbing stack vents are not dimensioned, roof drains may or may not be dimensioned. Mechanical equipment and penetrations are usually dimensioned on the mechanical roof plans.

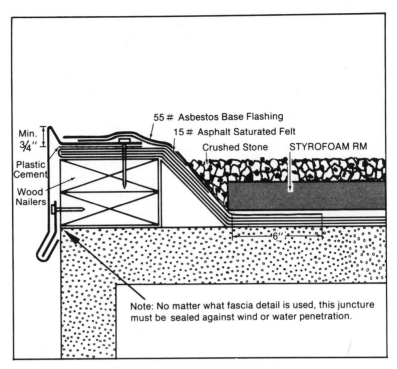

Figure 6-13. Gravel stop edge detail.

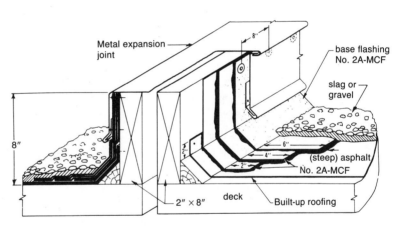

Figure 6-14. Expansion joint detail.

diffusers and registers, smoke detectors, sprinkler heads, access panels, and communication speakers (Fig. 6-18).

A reflected ceiling plan is best described as the reflection that would be seen in a mirror placed directly below the ceiling, hence the name, *reflected ceiling.* Every visible construction feature of the ceiling is shown, but nothing above the ceiling is included. For example, HVAC diffusers are shown but the connecting ductwork is not, that is shown on HVAC drawings. Similarily, sprinkler heads are shown but associated piping is not, that appears on fire protection drawings. Since the view is cut just below the ceiling, the doors, windows and other wall openings are not shown in the wall outline.

The reflected ceiling drawing is the result of the coordinated effort of the architect and engineers and will be used by contractors for the exact placement

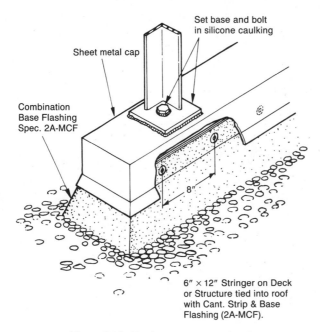

Set base and bolt
in silicone caulking

Sheet metal cap

Combination
Base Flashing
Spec. 2A-MCF

8"

6" × 12" Stringer on Deck
or Structure tied into roof
with Cant. Strip & Base
Flashing (2A-MCF).

Figure 6-15. Equipment support detail.

of their individual equipment. For example, the electrical drawings show the approximate location of lighting fixtures and the reflected ceiling plan establishes the exact position.

Architects develop the base reflected ceiling plan indicating the type and material of the ceiling and issue copies to the various engineering disciplines involved. Sometimes, this initial plan is developed in coordination with the engineering disciplines to ensure a compatable ceiling system.

As with other plan views, ceiling symbols should be included on the same sheet with the plan, space permitting; essential details may also be drawn on that same sheet. In cases where the plan sections and details must be located on other sheets, these sections and details should be placed in an organized manner so that they can be easily and rapidly located by the contractors. Re-

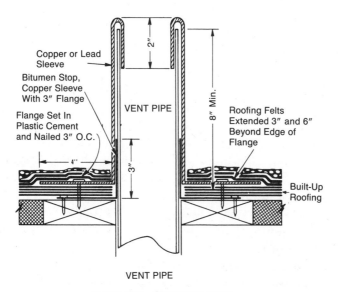

Copper or Lead
Sleeve

Bitumen Stop,
Copper Sleeve
With 3" Flange

Flange Set In
Plastic Cement
and Nailed 3" O.C.

2"

VENT PIPE

8" Min.

Roofing Felts
Extended 3" and 6"
Beyond Edge of
Flange

4"

3"

Built-Up
Roofing

VENT PIPE

Figure 6-16. Stack vent detail.

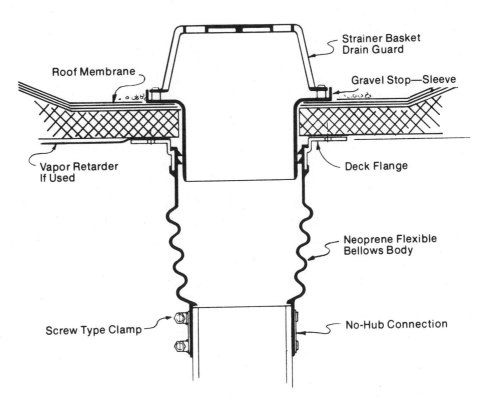

Figure 6-17. Roof drain detail.

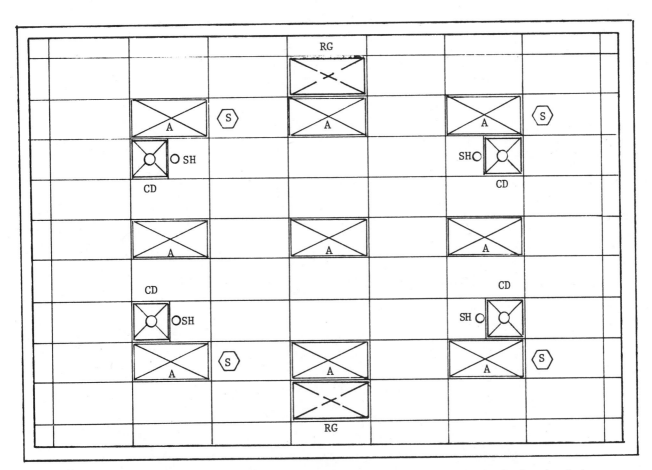

Figure 6-18. Reflected ceiling plans are drawn by architects with input from appropriate engineering disciplines so that their equipment can be fully coordinated and compatible with the ceiling style and grid. These drawings are generally not dimensioned but some equipment and device locations may be in special cases.

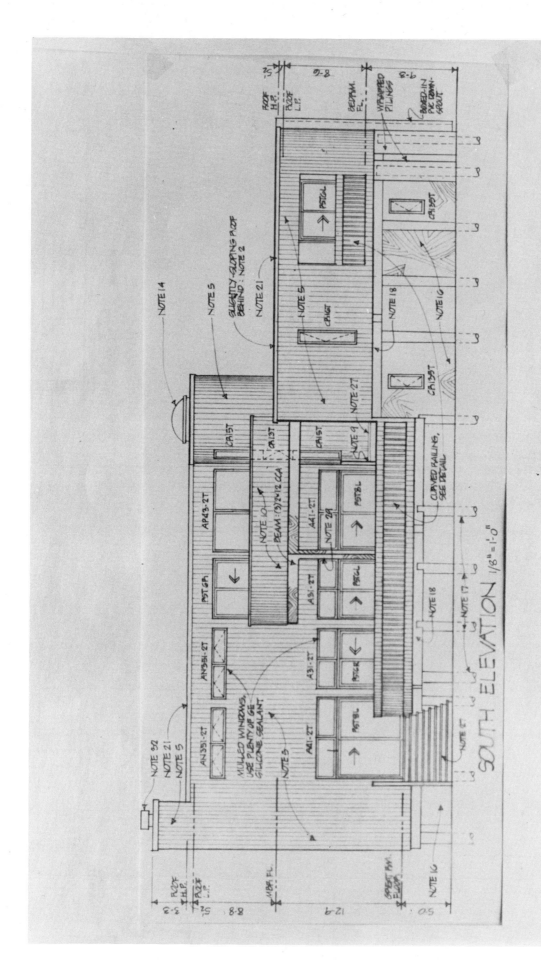

Figure 6-19. Exterior elevation of the south exposure is uncluttered by extensive lettering. The drawing illustrates the type of construction with notes used to describe the materials of construction. (Courtesy of Sidney Scott Smith, AIA, Architect. Reprinted by permission.)

106

flected ceiling plans are most frequently drawn at the same scale as the floor plan. All devices on the plan are identified by unique symbols and one or more letter designations, as appropriate. Where more than one type of lighting fixture, HVAC diffuser, etc., is used, different letters are needed and these are keyed to the symbols and/or descriptions in the notes, schedules, or specifications.

Reflected ceiling drawings require complete coordination among all disciplines involved to permit proper placement of the numerous devices involved. For example, the lighting fixtures must provide effective lighting distribution, ceiling diffusers must permit proper air throw and diffusion, sprinkler heads must satisfy spacing requirements and, at the same time provide a ceiling that *appears* uncluttered and symmetrical.

Where roof overhang is involved at that particular floor, the soffit is drawn and the devices are shown on the ceiling plan. Room names or numbers may be placed outside the plan proper with leaders indicating the space identified, or for single spaces, the identification may be placed under the drawing as, CONFERENCE ROOM REFLECTED CEILING PLAN.

6. ELEVATIONS

The term elevation is commonly applied to exterior views of a building. It is a vertical view and is identified by the exposure of the structure; for example, the south elevation is a drawing of the southern exposure or south face of the exterior wall as viewed by looking directly at it. Unlike the perspective drawing where objects further from the viewer are foreshortened, an elevation is an orthographic drawing. All dimensions are drawn to the same scale. Perspectives produce *pictorial drawings* that appear like actual photographs. Orthographs are graphic projections of buildings as viewed from an infinite distance.

Elevations are drawn at the same scale as the floor plan from which they are developed and when placed side by side on a drawing, adjacent elevations are placed together. For example, when the four elevations of a four-sided building are placed side by side, moving from one elevation to the other would present the same views as would be seen by walking around the building, i. e., adjacent corners are placed next to each other (Fig. 6-19, Fig. 6-20).

Elevations show the wall elements and materials, form of the roof where appropriate, and all wall openings. Windows and doors are identified by symbols related to the floor plan or by notes described in the window and door schedules. Overall horizontal and vertical dimensions, and intermediate dimensions to ceiling height, etc., are shown. Wall openings for doors, windows, wall louvers are not dimensioned on elevations; this relevant information should be on the associated floor plan. A grade reference line at the base indicating existing and new lines are shown. Hidden foundation work below grade may be indicated on the drawing by broken lines, particularly where step footings or other irregularities occur that cannot be adequately addressed on section cuts through the elevation. In cases where foundations are at a uniform elevation throughout and are detailed on sections, it is not necessary to redraw it on the elevations.

Elevations are frequently developed simultaneously with the sections through the elevations. This is necessary to coordinate the materials of construction with components such as doors, windows, and exterior heights. In the interest of economy of construction, full-sized coursing material is used and the related detail is usually best determined from a section taken through the elevation.

Interior elevations, sometimes called section elevations, are seldom identified

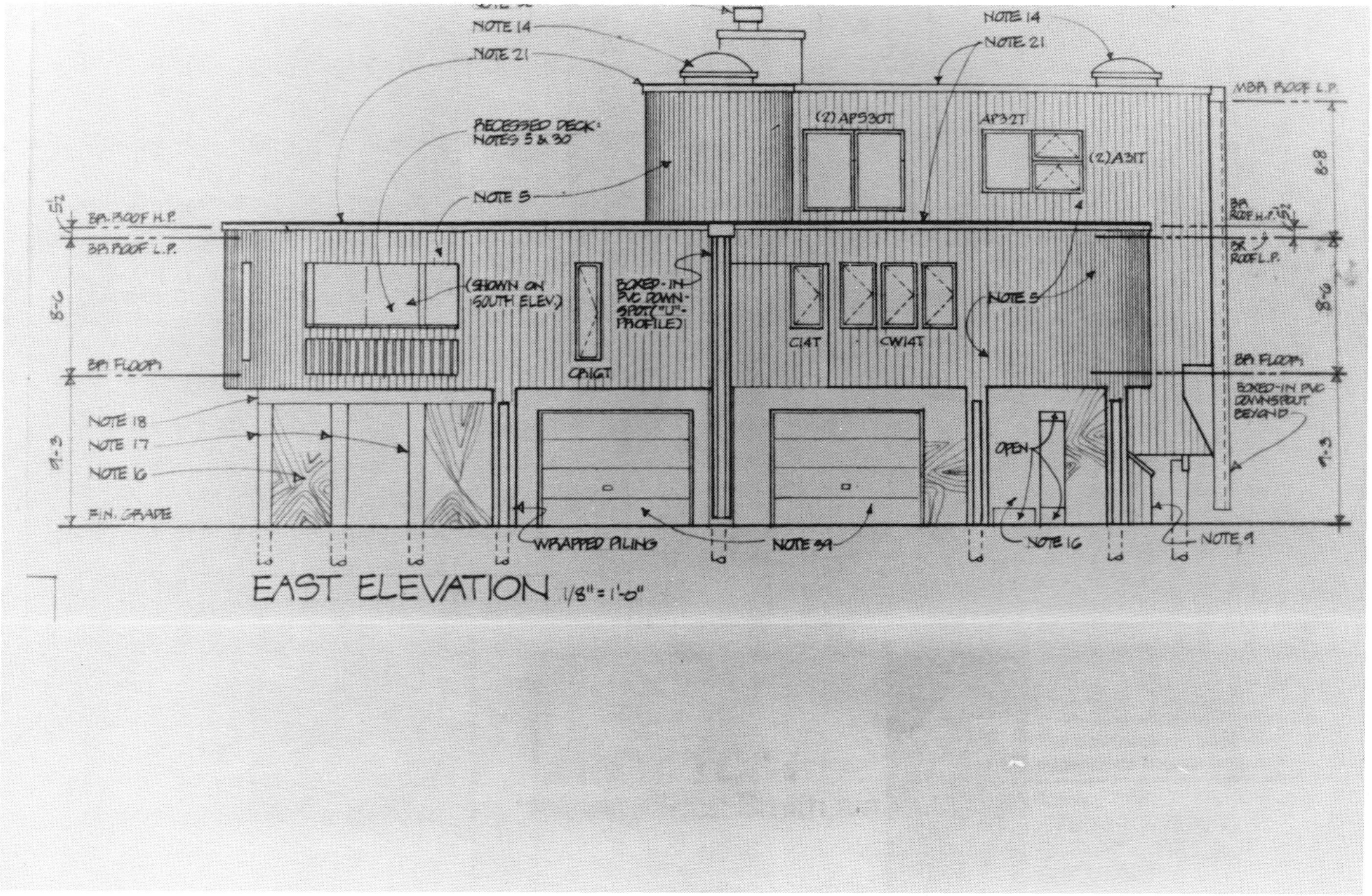

Figure 6-20. Exterior elevation of the east exposure of the same building in Fig. 6-19. (Courtesy of Sidney Scott Smith, AIA, Architect. Reprinted by permission.)

as *elevations;* instead they are described as to application. For example, when vertical information is necessary to describe a particular wall, an elevation of the entire wall is drawn and may be identified as SOUTH WALL OF CONFERENCE ROOM (Fig. 6-21).

Interior elevations are used to provide additional detail about interior features that cannot be adequately described on the floor plans. They are similar to exterior elevations, except that in most instances they show only floor-to-ceiling levels. Their primary function is to illustrate walls and portions of walls exactly as they would appear after construction. These drawings depict all visible features. Interior elevations serve two specific functions:

a. Provide adequate information to the contractor so that the project can be bid and eventually constructed.

b. Illustrate design intent to the client.

Floor plans show protrusions from, indentations into, and openings in walls. Floor plans are horizontal views of the floor and merely show walls as they appear in section. Elevations are vertical views between floor and ceiling lines and show all design and surface features of the wall. Interior elevations are not required and should not be drawn for walls that contain no surface mounted or built-in items or have no specific openings.

Interior elevations are usually drawn for kitchen, toilet, and special rooms. They are usually drawn at a larger scale than floor plans and exterior elevations. Interior elevations need not be drawn when the same information is described in room schedules.

Isometrics are often used to describe two adjacent interior room surfaces. Perspectives are used when three sides of a room can be better illustrated on one drawing of the interior space than with three separate elevations.

7. SECTIONS

There are several types of section drawings that are used to provide additional information that cannot be adequately described on plans and elevations. One type is cut through the building and shows interior spaces in elevation. Another is cut through walls to describe wall construction materials and methods (Fig. 6-22). Cuts are also made through windows and doors to provide the necessary details of construction and installation. These cuts are described as:

- Building cross sections
- Building longitudinal sections
- Wall sections
- Window and door sections
- Stair sections

Other than elevations, all drawings discussed so far have been horizontal views; these includes site plans, floor plans, reflected ceiling plans, and roof plans. Sections are vertical views slicing the building or wall. The best example of a section is to take, for example, a birthday cake and slice it vertically through the middle. The exposed piece is a section through the cake. The same principle applies to all section cuts.

Building sections are usually drawn at ½ inch scale. Wall, door, openings, and stair sections are drawn at larger scales of 1 inch or 1½ inch depending upon the degree of detailing needed to show the important elements.

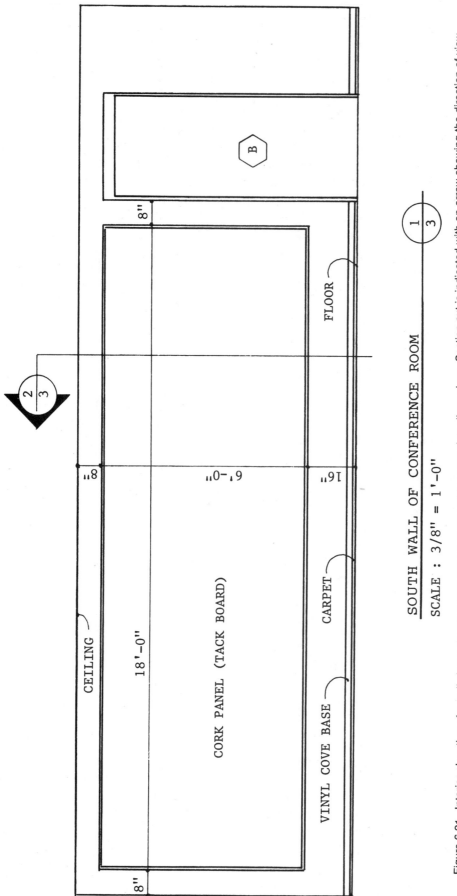

SOUTH WALL OF CONFERENCE ROOM

SCALE : 3/8" = 1'-0"

Figure 6-21. Interior elevation of a wall shows the view from floor to ceiling with appropriate dimensions. Section cut is indicated with an arrow showing the direction of view.

CEILING

18'-0"

CORK PANEL (TACK BOARD)

VINYL COVE BASE

CARPET

FLOOR

B

8"

8"

8"

6'-0"

16"

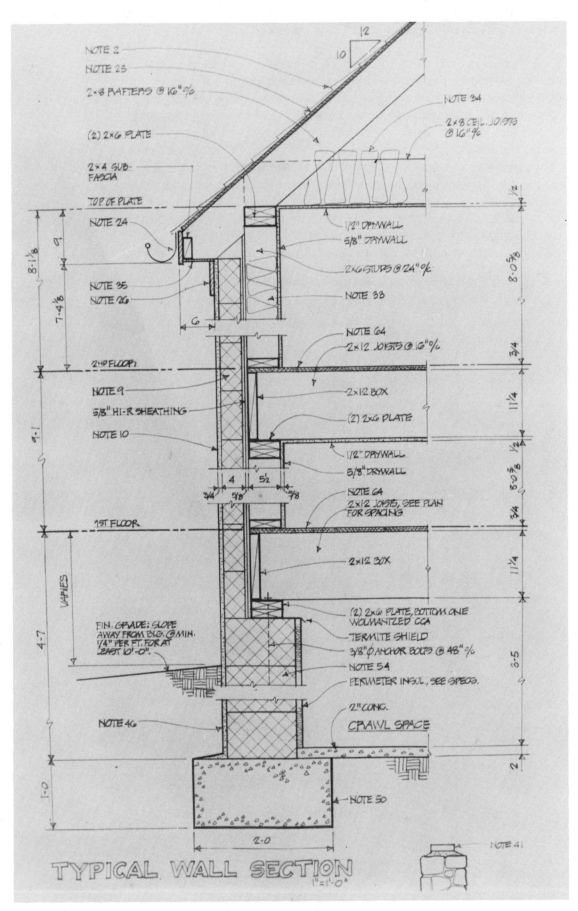

Figure 6-22. Wall section clearly describes wall construction method and graphically illustrates (by symbols) type of construction materials. All vertical dimensions are shown and floor levels are identified. Neatly aligned notes, away from the wall section with leaders to the point of application, yield a clear, uncluttered drawing, and simplify drafting and reduce drafting time. (Courtesy of Sidney Scott Smith, AIA, Architect. Reprinted by permission.)

Building sections are also used to illustrate surface mounted or built-in features on interior partitions. Frequently, additional sections are cut through these sections to provide additional information. Unfortunately, some section cuts are noted as details and this may cause some confusion to entry-level drafters.

Sections are essential not only to building contractors for bidding and construction work, but also to drafters so that they can check to make sure that all elements fit in place as intended by design. Often the third dimensional investigation by the drafter may indicate that revisions or dimensional adjustments must be made to previously completed plans or elevations to make things work in actual construction. The number of sections and details developed depends upon the complexity of the building components.

Figure 6-23 is a section cut through the wall elevation shown in Figure 6-21.

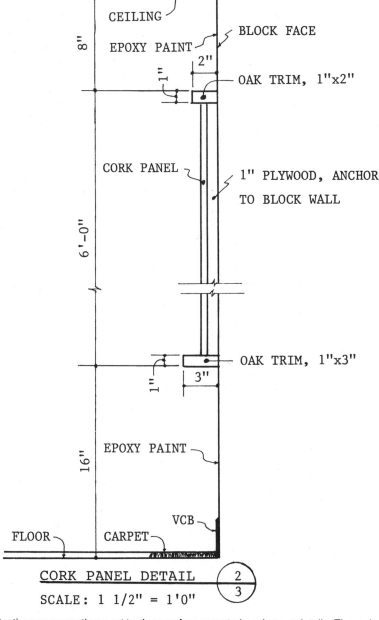

Figure 6-23. Sections are sometimes cut to show surface mounted equipment details. These do not generally show complete wall construction details.

It provides the necessary construction details for the installation of the cork panel. In some instances, elevations are used with appropriate notes for simple construction methods, but in cases where the item to be installed is field constructed, it is better to show by a section the intended materials and methods.

8. DETAILS

Details are enlarged drawings that provide essential specific information. They are used to describe and define areas that require additional emphasis. The best way to visualize a detail drawing is as a close-up photograph. Take for example, a typical window installation:

1. The plan shows the window's horizontal location and its dimensions; it is keyed to the window schedule by a symbol (Fig. 6-24).
2. The elevation then provides additional information, such as appearance and vertical dimensions. This view should also use the symbol for identification.
3. The section through an elevation at a window shows still another picture. (Fig. 6-25). It shows, generally, in simple form the window relationship to the wall and more specific dimensions. Details are usually not provided for standard manufactured windows in simple wall construction. When the wall construction at the window sill jamb and the head is unusual, a detail is required.
4. This area is usually circled up on the section view and that portion of the construction is enlarged sufficiently to describe every minute detail, hence the name, *detail.* (Fig. 6-26).

Details are also provided for areas that are too small on plans to describe fully and dimension accurately. These areas include toilet rooms, locker rooms, kitchens, stairs, mechanical rooms, etc. The areas are usually identified on the plan views (or elevations) by a simple note *SEE DETAIL A-7.* These details are generally combined on the sheet entitled *DETAILS*

Detail drawings are drawn at 1 inch or 1½ inch scale, or even larger if necessary. This information is primarily for the contractor's benefit. Most contractor's questions, possible errors, and construction delays are the result of poorly detailed drawings. A well-designed building with all required details reduces the total involvement of the architect during the construction phase. Details may be cut or indicated on plans, elevations, and sections (Fig. 6-27). Some of the more common areas where detailing should be used are as follows:

1. Building framing connections, including columns, beams, joists, and walls.
2. Roof openings and terminations at walls showing flashing details, gravel stops, and cant strips.
3. Poured concrete foundations, floor and footing connections showing reinforcement, water stop insertion, sealant application, vapor barrier and insulation attachment, and expansion joints around columns.
4. Stairs to describe framing, connections to stair wells, riser and run dimensions, and railings and its method of attachment.
5. Door and window sill and lintel installation.

Details are identified similar to sections cuts. When detail drawings are not in close proximity to their cut location where a leader can be used to connect the detail to the cut, the two-number system is used. To repeat, the top number

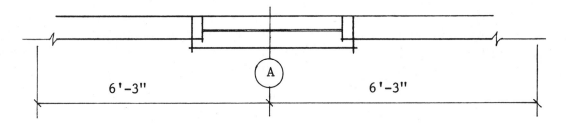

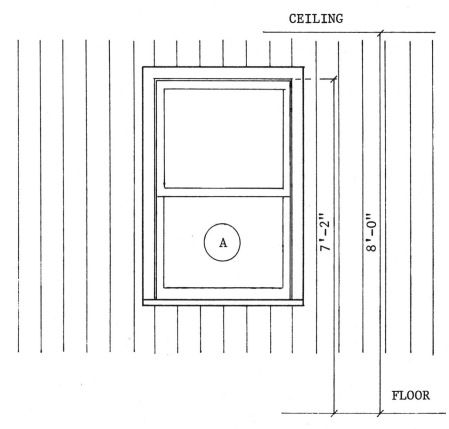

Figure 6-24. Typical plan and elevation views of a double hung window in a frame wall with vertical siding indicated on the elevation view. Note method for dimensioning in both cases. The letter "A" in the circle symbol refers to the type of window described in the schedule.

identifies the detail and the lower number identifies the drawing sheet where the detail appears (Fig. 6-28).

Details are sometimes identified by circling up the specific area intended to be enlarged. This circle is then identified by a leader and the numbers. Where details are cut through another drawing, the cut and direction of view are indicated.

Some firms, depending upon the scope of the detailing required, frequently cluster all details on one drawing identified as the detail drawing. On small and less complex projects, it is more common to place the detail on the same sheet where it applies.

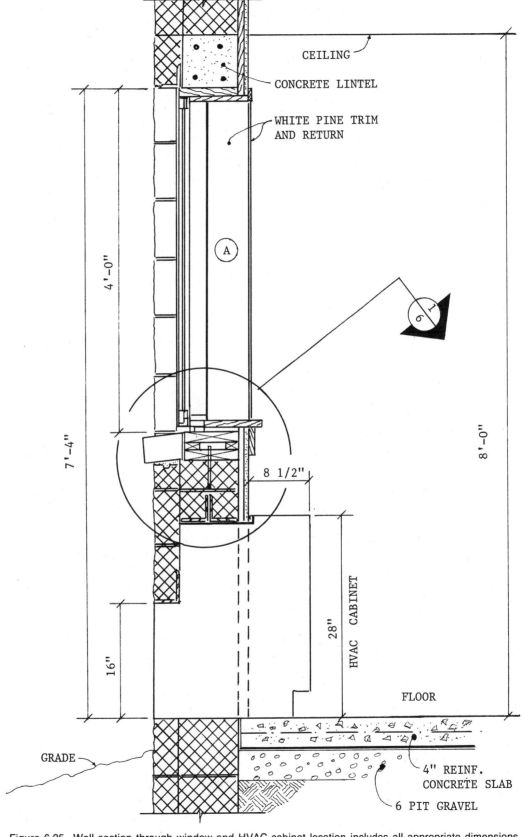

Figure 6-25. Wall section through window and HVAC cabinet location includes all appropriate dimensions and identifies construction material by typical construction material symbols. Circled up area indicates that there is detail No. 1 on Sheet No. 6.

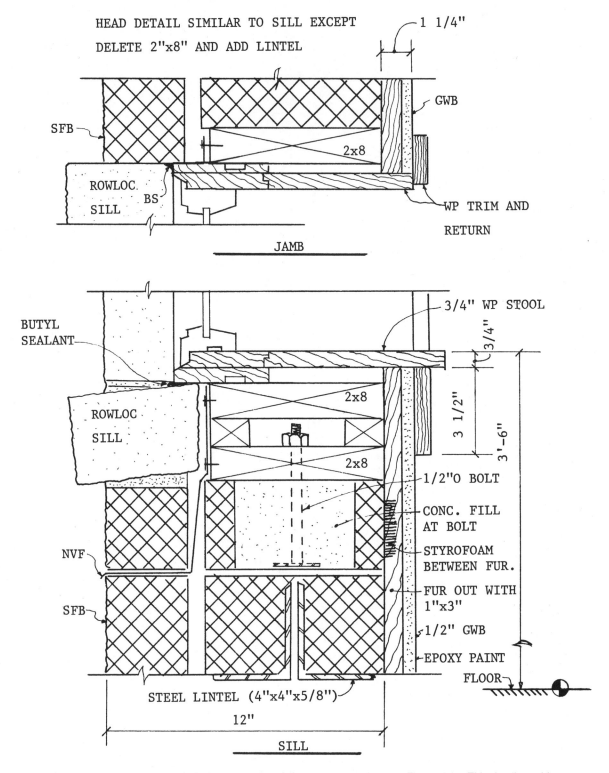

Figure 6-26. Enlarged detail of the wall section indicated by circled cut on Figure 6-25. This detail provides additional information for the sill and jamb construction.

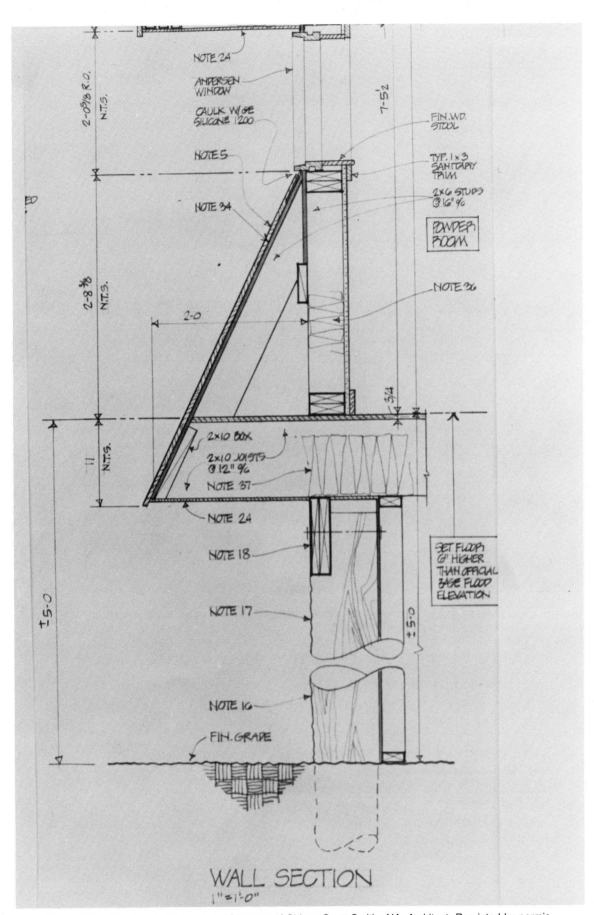

WALL SECTION
1"=1'-0"

Figure 6-27. Framing detail drawing. (Courtesy of Sidney Scott Smith, AIA, Architect. Reprinted by permission.)

SYMBOLS   DESCRIPTIONS

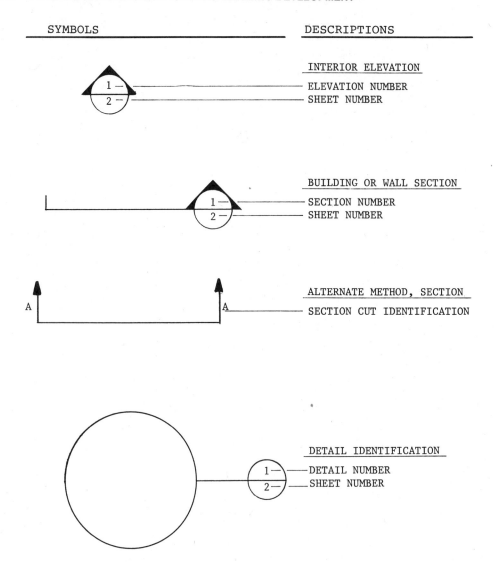

INTERIOR ELEVATION
ELEVATION NUMBER
SHEET NUMBER

BUILDING OR WALL SECTION
SECTION NUMBER
SHEET NUMBER

ALTERNATE METHOD, SECTION
SECTION CUT IDENTIFICATION

DETAIL IDENTIFICATION
DETAIL NUMBER
SHEET NUMBER

Figure 6-28. Typical symbols used to identify interior elevations, section cuts, and detail areas. The top symbol is used for interior elevations and the arrow indicates view direction. The center two symbols indicate location and view direction of building cross section and wall section cuts. The bottom symbol is used circle up area enlarged for greater detail, may be used with or without an arrow, as appropriate.

In the past, and to some extent even today, many manufacturers include various details in their literature that drafters merely have to trace onto their drawings. More and more manufacturers today provide these details in adhesive-backed film or vellum that drafters merely apply to their drawings.

Many offices design their own typical details of components that are frequently used. These are also on adhesive-backed transparencies that can be stuck to the drawing. With this technique the drafter merely letters in the identification and the detail is completed. This practice is more common in engineering design than architecture. Engineers have more items that are repeated more frequently on various building design.

9. SCHEDULES

Schedules are shorthand techniques used by architects and engineers to simplify drafting and reduce drafting time. Schedules are a tabulation of common items

NO.	WIDTH	HEIGHT	TYPE	MATERIAL	FINISH	FRAME	GLASS	REMARKS
1	3'-0"	7'-0"	A	WOOD	NATURAL	HM	NONE	- - -
2.	2'-0"	7'-0"	C	WOOD	PAINTED	HM	NONE	- - -

DOOR SCHEDULE

Figure 6-29. Door schedules show the door symbol used and lists all pertinent information.

required on the project that have similar characteristics. Whenever three or more similar items are required on a building, schedules should be used. When this information is properly organized in schedules, much of the detail lettering on drawings is eliminated. Since building construction drawings are graphical presentations of the project, they should not be unduly cluttered with notes.

Architectural schedules are usually developed for doors, windows, room finish, hardware, kitchen equipment, lintels, access panels, fire extinguishers. Additional schedules may be required depending upon the particular building type and the requirements of construction (Fig. 6-29).

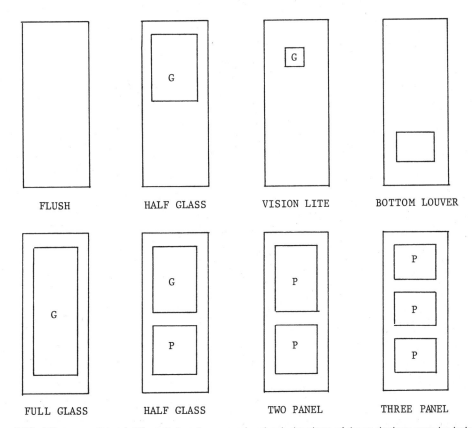

Figure 6-30. When a variety of different door types are involved, drawings of these designs may be helpful. The different designs may be identified by a letter symbol on the door and so noted on the schedules.

ROOM FINISH SCHEDULE			WALLS				CEILING	REMARKS
NO.	FLOOR	BASE	NORTH	EAST	SOUTH	WEST		
1	V.A. TILE	VINYL	PTD. A	PTD. A	PTD. B	PTD. A	ACC. TILE	– – – –
2	CARPET	VINYL	PTD. A	PTD. A	PTD. B	PTD. A	ACC. TILE	– – – –
3	CT	CT	CT	CT	CT	CT	2 HR. DW.*	* PTD. EPOXY

Figure 6-31. A typical room finish schedule. The type of material used is fully described in the specifications. The letters ''A'' and ''B'' identifies the paint color.

Schedules should be prominently titled with larger letters than those used in the body of the schedule. The tabular format contains horizontal and vertical line separations and the entire schedule is boxed. The first left-hand column contains the schedule equipment identification number, and what follows depends upon the particular schedule. The far right-hand column is used for notes. The symbol used for the schedule should be included adjacent to the title of the schedule (Fig. 6-31).

Horizontal line spacing of ½ inch is usually adequate since this permits two lines of ⅛ inch lettered notes, if needed. Column spacing of ¾ inch is commonly used but should be wide enough to permit adequate spacing for the text. Abbreviations may be used on schedules providing that they are defined under the list of abbreviations. Lettering on schedules must be as neat and distinct as all other lettering on the drawings. Schedules are a useful shorthand technique and should be considered whenever practical.

PART III
ENGINEERING DRAFTING AND DRAWING DEVELOPMENT

Chapter 7
Introduction

Engineering disciplines in an A/E office usually match the building trades in construction. For example, electrical engineers develop drawings that will be used by the electrical contractor to perform his work. Electrical engineers also produce the electrical section of the specifications related to the electrical drawings. The same method applies to other engineering disciplines and building construction trades.

This and what follows does not intend to define building trade jurisdiction, i.e., establish responsibility of work for the various trades. Today, in this era of highly technical specialization, there may even be subdivisions within the various building trades.

Generally, architectural drawings are used by the general contractor (GC) who may act as a broker and subcontract all of the work or may perform some and subcontract some. On single prime (general) contract projects, the GC will subcontract specific trades such as mechanical, electrical, etc. On multiple separate-trade contracts, which are mostly used on public works projects, the various trades bid separately and each perform its own work.

While there is some overlap between architectural and engineering disciplines, and this depends upon size of the A/E firm and its internal structure; engineering disciplines are mostly totally separated from architectural disciplines.

In small firms and on small projects, architects may perform the basic architectural work plus the building structure and in many cases may even do the site work. In large firms and on large projects, the structural design and drawings are performed by engineers. Site work may include input from civil, mechanical, electrical, plumbing, and fire protection. The drawing may be identified by a prefix that indicates the various trades involved. This prefix may include one or more letters, such as E, M, P, FP, as required. This indicates to the contractor exactly what work is included on the drawing.

Generally, the engineering disciplines are broken down as follows:

ENGINEERING DISCIPLINES

1. Structural
   a. Building structure
   b. Civil (site)
   c. Soils consultants
   d. Testing services
2. Electrical
   a. Power and light

    b. Security
    c. Instrumentation
    d. Communications
    e. Lighting consultants
    f. Lightning consultants
3. Mechanical
    a. Steam and power
    b. HVAC
    c. Automation
    d. Mechanical acoustics
    e. Testing and balancing
4. Plumbing
    a. Potable water distribution
    b. Sanitary drainage
    c. Storm drainage
5. Fire protection
    a. Sprinkler system
    b. Life safety

Again, this breakdown depends upon the type and size of the A/E firm and the size and complexity of the building construction project. For example, on small projects the mechanical contractor may jurisdictionally perform: heating, ventilating and air conditioning; automation; plumbing; and fire protection.

Basically, engineering disciplines are subdivided according to the various building trades involved. Generally, drafters, designers, and engineers usually perform in a specific discipline. Each engineering discipline is highly specialized and requires unique skills to perform within a specialized discipline.

Engineering drawings are not stand-alone drawings. Engineering work is indicated on basic architectural building outline drawings and therefore this drafting cannot be performed until architectural plans are developed. While architectural drawings are fully dimensioned so that the structure can be constructed precisely as shown, engineering drawings are not dimensioned in the same sense.

Architects design the building; engineers provide the necessary environmental information such as HVAC, electrical, plumbing, etc., to make the structure habitable. Engineering drawings show, by discipline, equipment, controls, devices, and connections.

Engineering drafting, while similar to architectural drafting, requires a somewhat different approach. First, and most importantly, buildings are constructed from architectural, and/or structural drawings, depending upon the scope of the project. On smaller projects, the structural work may be included on the architectural drawings, and all building construction is performed from these drawings. Architectural drafting is required to be more precise and accurately dimensioned, and architectural drawings show all necessary building construction sections and details. Engineering drawings do not show building sections and details, and other essential construction information.

Engineering drawings, with the exception of structural, are graphical in nature and in many instances piping and duct runs are indicated by single lines in the general location. HVAC ductwork is dimensioned for size but not for location. Electrical and mechanical equipment on floor and roof plans are not dimensioned. The exact location will be indicated on architectural and/or structural drawings. Where dimensioning and exact location are necessary, larger scale drawings are made for those specific areas.

Engineering drawings are made by several different design disciplines, depending upon the size of the firm and the manner in which it is established. In larger firms, each different discipline is headed up by a chief engineer and the department includes project engineers, designers, and drafters. In small firms, sometimes several disciplines are combined under one chief engineer, or in some cases, the owner of the firm. For purposes of definition, engineering disciplines are generally divided into electrical, mechanical, plumbing, fire protection, and structural.

Depending upon office practice, engineering may or may not require making floor plans. Some architects provide what is known as transparent, erasable polyester film from the original architectural drawings. Erasable films are reverse printed with lighter line weight building outlines to permit prominent illustration of engineering work, yet reproducible for building outline background. These are given to the various engineering disciplines. Electrical usually receive two floor plans, one each for lighting and power. Mechanical, plumbing, and fire protection usually get one each. With this technique the building outline and interior walls and partitions are identical on all architectural and engineering drawings. These reproducibles are made without any architectural dimensions, architectural equipment and features, room identifying names or numbers, and may or may not show doors.

In larger firms, each chief engineer has one or more project engineers, who, as the title implies, is responsible for one or more projects. As a licensed professional engineer, project engineers have technical responsibility for the project. The amount of responsibility delegated by the project engineer depends upon the qualifications of the designers and drafters. Designers may be involved in various technical aspects of the project such as system design, equipment selection, and specifications. Drafters seldom have technical responsibilities but must know systems and equipment and be able to layout drawings, place equipment accurately, develop interconnections, and must be familiar with the use of appropriate symbols because drafters make the drawings.

Some firms provide the engineers with prints made from the architects drawing. The engineer drafter will have to, at least, trace the essential information from the architect's prints. The building outline should be traced with lighter line work than will be used for the equipment. In neither case is the engineering drafter required to do detailed building work.

This is not intended to imply that engineering drafting is simple, to the contrary, the unique requirements of the various engineering disciplines are sufficiently different so that in all but the smallest engineering offices, drafters work in only one discipline. Drafters, and this applies also to designers, may specialize in structural, electrical, HVAC, plumbing or fire protection. This specialization is also to the drafter's advantage. Advancement within a specific discipline depends upon the attainment of a particular level of proficiency. This proficiency is easier to acquire by working in one discipline.

While architects may provide either transparencies or prints of floor plans to the engineers, there are other drawings engineers must make. Electrical, HVAC, plumbing, and fire protection drafters are generally required to draw roof plans, enlarged plans of rooms where their specific work is involved, show all equipment and interconnecting piping, ductwork, etc., plumbing riser diagrams, electrical single-line diagrams, and numerous details showing how equipment is installed, mounted, attached, hung, and interconnected. Depending upon the scope of the project, these could be very involved.

Structural drafting is different from other engineering disciplines. For example, structural does not show floor plans as such, but shows the structural support for the floors and roof. Essentially, structural drawings show the structural skeleton of the building and appropriate details necessary for the erection of the structure. The structure of the building must be fully coordinated with the architecture to ensure compatability.

The best approach to the discussion of engineering drawings and engineering drafting is by addressing each different discipline individually.

Chapter 8
Electrical

Electrical drawings usually consist of two floor plans, one for lighting and one for power. On small projects and for residences, one is usually sufficient. Roof plans may or may not be needed depending upon what electrical and mechanical equipment requiring power is installed. Projects that have roof-mounted air conditioners, electrical transformers, exhaust fans, lightning protection equipment, communication equipment, will certainly require a roof plan. Electrical drawings are required for single-line power distribution diagrams, schedules, and equipment installation details.

1. FLOOR PLANS—LIGHTING

Electrical lighting plans are generally not stand-alone documents. The number and general location of lighting fixtures should have been coordinated with the architect, particularly if a dropped (false) ceiling is being used. Then the architectural reflected ceiling plan is used for the ceiling installation and the exact location of all fixtures. Fixtures on electrical drawings are scaled but are not dimensioned. The quantity and type are shown on electrical drawings and the type may be described in notes, in schedules on drawings or in the electrical section of the specifications.

Electrical lighting plans also show the type and location of switches used to control the fixtures and frequently identify the grouping of fixtures by simple curvilinear lines between fixtures and from fixtures to switch(es). Generally, a short line from the switch (or lighting group) with an arrow indicates a conduit and/or cable continued to the panel. The circuit and panel number is indicated at the end of the arrow.

Conduit and cable runs are graphical in nature and not intended to show the exact location of the run. The contractor is given the latitude to make lighting fixture interconnects and home runs to panels in the most economical manner that can be accomplished, consistent with the requirements of the National Electrical Code, and coordinated with all other building trades.

When several different types of fixtures are used on the project, they are described on the symbol and legend schedule and only the symbol is used on the drawings.

The most important aspect of lighting plan drafting is the knowledge of symbols required for the fixture and switches, and how to draw the equipment and devices properly. The fixture sizes must be proportioned to the scale of the drawing. For example, a 2 × 4 foot fluorescent fixture on a scale of ¼″ = 1′ -0″ will be drawn as a ½ inch by 2 inch rectangle. This rectangle may contain letter symbols representing the types of fixtures when more than one of the size shown is used. Proportional scaling is essential in order to illustrate fixture relationship to the space properly.

Many A/E firms have their own symbol list, while others may use national standards because these are more universally recognized. The drafter needs to know the symbols before attempting to draw electrical lighting plans.

Figure 8-1 is a lighting floor plan of one room of a building. It shows some of the devices usually found on these plans, not necessarily in the quantity and type shown here. Since there is only one lighting panel in the building it is not numbered. Also, the wiring home runs noted by the arrows indicate the circuit number in that panel. When more than one panel is necessary, each panel is identified as, LP 1, LP 2, and so on, and the home runs then indicate the circuit number of the specific panel to which the home run is connected as LP 1—1, LP 1— 2, LP 1—3, etc. The number of circuit wires required are shown by slash marks on the home runs; ground wires are not included in the number but are required by code.

In order for the contractor to know how the fixtures must be circuited and switched, this information must be shown. For example, three fixtures are shown wired to 3-way switches located at each main entrance. The other six are controlled by a single-pole switch. Three wall washer fixtures are switched as a group by a single-pole switch whose location is indicated.

Exit lights are wired independently and the circuiting is usually described on the single-line diagram. Clock outlets may be shown on either the lighting plan or the power plan. Fluorescent fixtures are shown to scale and the location is coordinated with the reflected ceiling plan. The letter A identifies the type of fixture with a full description provided on a fixture schedule, in notes, or in the specifications. Since there could be three types of panels used on electrical drawings, symbols in addition to letter notations are used to identify each. Typically there could be lighting panels, power panels and telephone panels (cabinets). See Figures 8-2 and 8-3 for commonly used symbols.

Home runs are never shown continued to the panel since that would serve no useful purpose and would unnecessarily clutter an otherwise clean drawing. Engineering drawings are generally diagrammatical in nature; their purpose is to show the types of equipment and devices and their general location. For example, the three switches shown on the drawing at one door seem to be spread over a two-foot space while actually they are usually installed in one three-gang box.

Figure 8-2 is an electrical lighting symbol list for the drawing in Figure 8-1. In actual practice, the electrical symbol list usually combines all the symbols on one list that are used on all the electrical drawings.

Figure 8-3 is a symbol list used by one engineering office. It is prepared on an adhesive-backed transparent film that is pasted on the appropriate drawing of the set. It includes most of the symbols generally used on a medium-sized commercial project. When additional, special symbols are needed, these can be readily added to the bottom of the list. The use of these adhesive-backed transparencies greatly simplifies drafting and saves time; since the description is typed, the end result is also more legible.

Figure 8-4 is a partial lighting floor plan showing different sized fluorescent fixtures readily identifiable as 2 × 4 and 4 × 4 sizes. Also note the darkened 2 ·× 4s in the drawing; this is used to indicate night lighting, and could have been accomplished by letter symbols but shading produces greater emphasis. Elec-

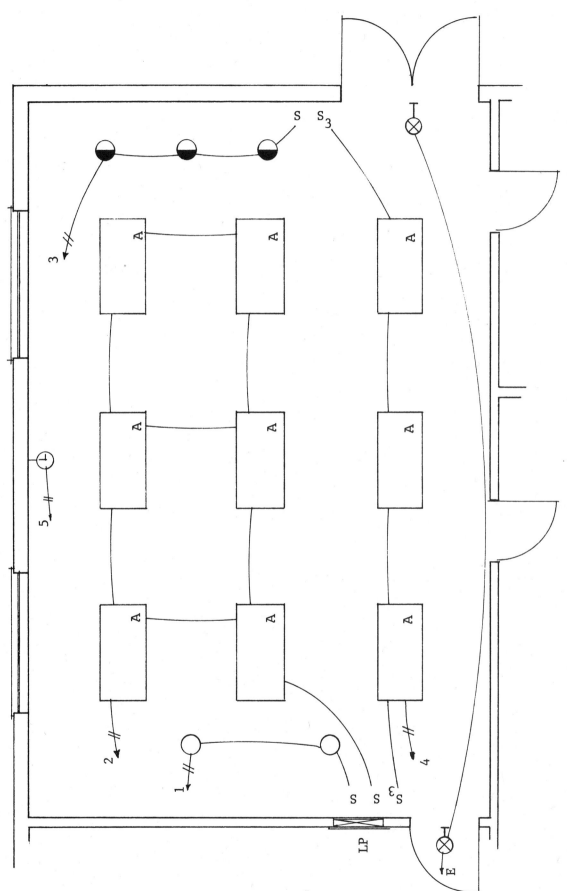

Figure 8-1. Lighting floor plan for one room of a building. Fluorescent lighting fixtures are identified by the letter "A" since other types of lighting fixtures are used elsewhere in the building. Complete fixture description is provided under the fixture schedule, in notes, or in the specifications. The number of wires and the circuit numbers are shown on home runs, indicated by arrows, to the panel. The panel number is omitted since only one LP is required.

ELECTRICAL SYMBOL LIST - LIGHTING

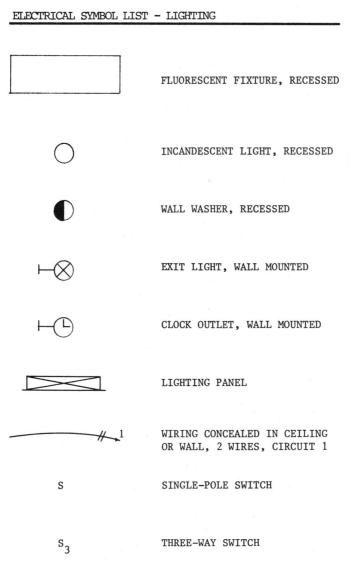

FLUORESCENT FIXTURE, RECESSED

INCANDESCENT LIGHT, RECESSED

WALL WASHER, RECESSED

EXIT LIGHT, WALL MOUNTED

CLOCK OUTLET, WALL MOUNTED

LIGHTING PANEL

WIRING CONCEALED IN CEILING
OR WALL, 2 WIRES, CIRCUIT 1

SINGLE-POLE SWITCH

THREE-WAY SWITCH

Figures 8-2. Symbol list for equipment and devices shown on the lighting floor plan in Figure 8-1. (From ANSI Y32.9-1972)

trical floor plans are not dimensioned; this was detailed on the architectural drawings. As for the location of the electrical fixtures, this is determined from the reflected ceiling plan. All that is necessary on lighting floor plans is the general location and the type of fixtures, specified by appropriate symbols and letters (or shading), switching, circuiting, and interconnecting wiring. While this may sound simple and easy to do, great care must be exercised to properly illustrate and identify all devices and equipment required by that plan. Drafters and designers must be able to draw neatly and legibly, with speed and accuracy.

2. FLOOR PLANS—POWER

Electrical power plans are quite different from lighting plans and may be rather involved, depending upon the scope and complexity of the project. Lighting plans show equipment and devices that are purchased and installed by electrical con-

SYMBOL LIST

	FLUORESCENT LIGHTING FIXTURE
	EMERGENCY FLUORESCENT LIGHTING FIXTURE
	CEILING OR WALL MOUNTED INCANDESCENT OR MERCURY FIXTURE
	EMERGENCY INCANDESCENT OR MERCURY FIXTURE
	EXIT LIGHT MOUNTING AND ARROWS AS INDICATED
S	SINGLE POLE SWITCH
S₃	THREE WAY SWITCH
Sᴘ	SWITCH WITH PILOT LIGHT
Sм	MANUAL STARTER WITH THERMAL OVERLOAD AND PILOT LIGHT
Sₐ	TYPICAL SWITCH DESIGNATION. INDICATES CONTROL OF SPECIFIC OUTLET
	DUPLEX RECEPTACLE
	FLOOR OUTLET
	SOLID CONNECTION
(P.S.) (F)	TELEPHONE OUTLET (P.S. - PAY STATION. F - FLOOR OUTLET)
	PANELBOARD
	JUNCTION BOX
	MOTOR OUTLET
	DISCONNECT SWITCH
	FUSED DISCONNECT SWITCH
	COMBINATION MOTOR STARTER
	HOME RUN
	WIRING CONCEALED ABOVE
	WIRING CONCEALED BELOW
	WIRING EXPOSED
E	EMERGENCY WIRING
	FIRE ALARM PULL STATION
	FIRE ALARM HORN
	FIRE ALARM STATION WITH HORN & FLASHING LIGHT ABOVE
	FIRE ALARM FLASHING LIGHT
AFF	ABOVE FINISHED FLOOR
WP	WEATHERPROOF
EWC	ELECTRIC WATER COOLER
GFI	GROUND FAULT INTERRUPTER
+ S +	(+) INDICATES DEVICE MOUNTED ABOVE COUNTER
	CLOCK OUTLET AND SINGLE FACE CLOCK

Figure 8-3. Typical actual electrical symbol list combines both lighting and power symbols.

tractors, whereas power plans must show electrical connections and circuiting for equipment purchased by the electrical contractor, plus any other equipment that may be specified by the architect, HVAC, plumbing, fire protection, kitchen equipment supplier, etc. Electrical power plans must be coordinated with all other design disciplines having equipment requiring electrical connections.

The following is a list of some equipment that may be found in a building requiring electrical connections, and the particular disciplines involved:

Architect

• Elevators
• Escallators
• Kitchen equipment
• Medical equipment

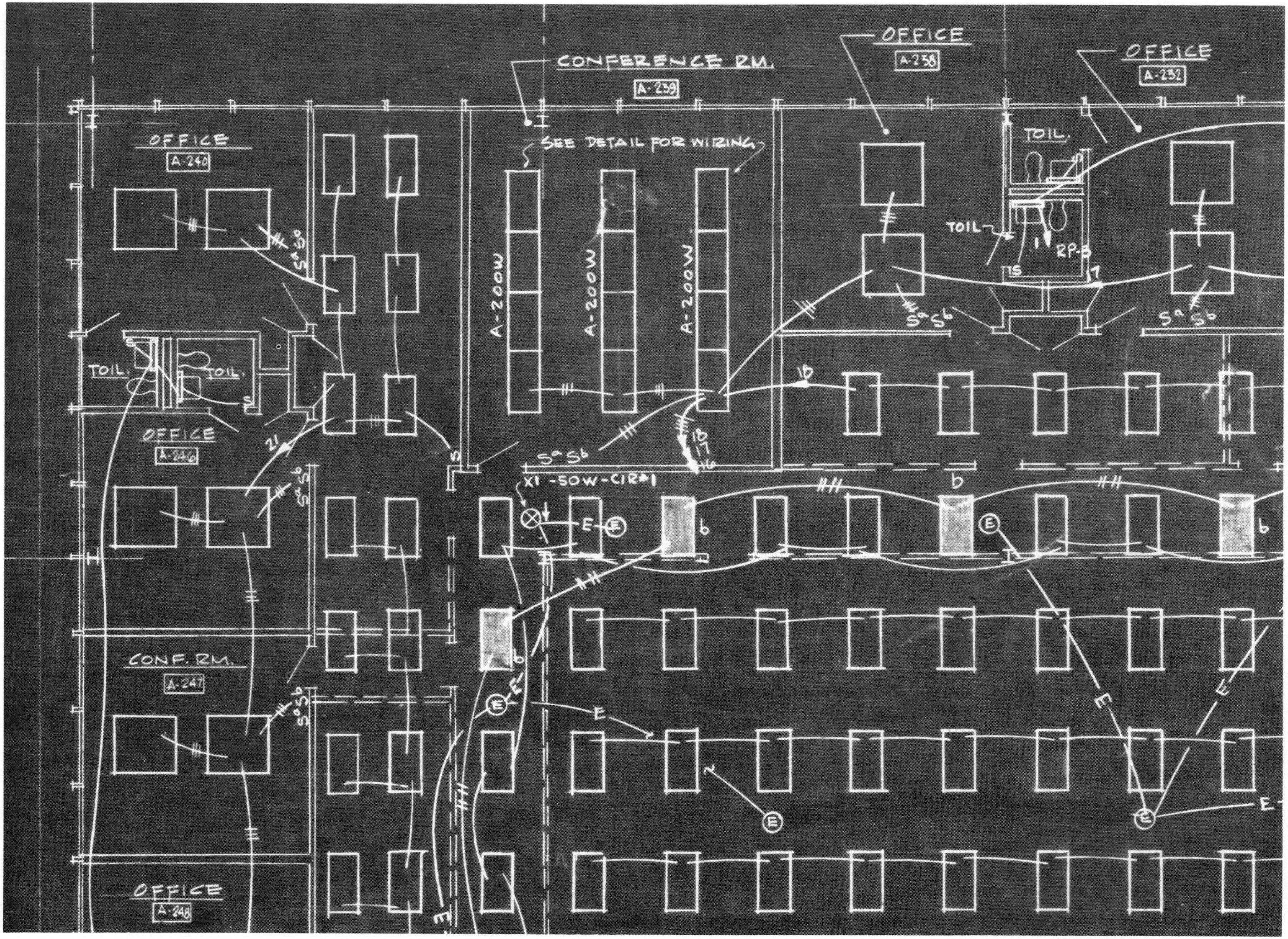

Figure 8-4. Actual office building lighting plan for a portion of one floor.

HVAC

- Air conditioning chiller
- Air handlers
- Cooling towers
- HVAC pumps
- Exhaust fans
- Boilers
- Electric duct heaters
- Motorized dampers
- Electric steam generators
- Air compressors

Plumbing

- Electric coolers
- Sump pumps
- Circulating pumps
- Sewage ejectors
- Electric hot water heaters

Fire protection

- Fire pumps
- Jockey pumps
- Fuel oil pumps
- Engine driven boosters

One of the greatest concerns of electrical drafters and designers is to not miss equipment requiring electrical connections. The second greatest concern is to show the equipment requiring electrical service correctly and in the correct location. It is for these reason that electrical power drawings must be fully coordinated with all other design disciplines and be continually monitored to pick up any design changes by the other disciplines. Last minute changes in equipment size, type, or location must be identified and appropriate corrections must be made to eliminate design errors. A common practice in some A.E firms is to hold regularly scheduled coordination meetings for the exchange of pertinent design information. Design disciplines are required to make progress prints at regular intervals during the working drawing phase for distribution to all other disciplines to ensure fully coordinated set of drawings

Power floor plans show all electrical and other discipline equipment and devices found in that space. Home runs are indicated by properly identified arrows and the number of wires involved are indicated by crosshatch marks. Equipment such as exhaust fans, copying machines, and electric water coolers are shown on the drawings in their proper locations with the required electrical service. The exhaust fan is described in the equipment schedule, in a note or in the specifications, by the mechanical discipline. The electrical requirements for the fan may be noted at the fan, in a fan schedule, or in the electrical section of the specifications. A simple note at the fan may suffice, i. e., "¼ hp, 120 v, 1-phase, 60 Hz, 4.2. amps" (Fig. 8-5).

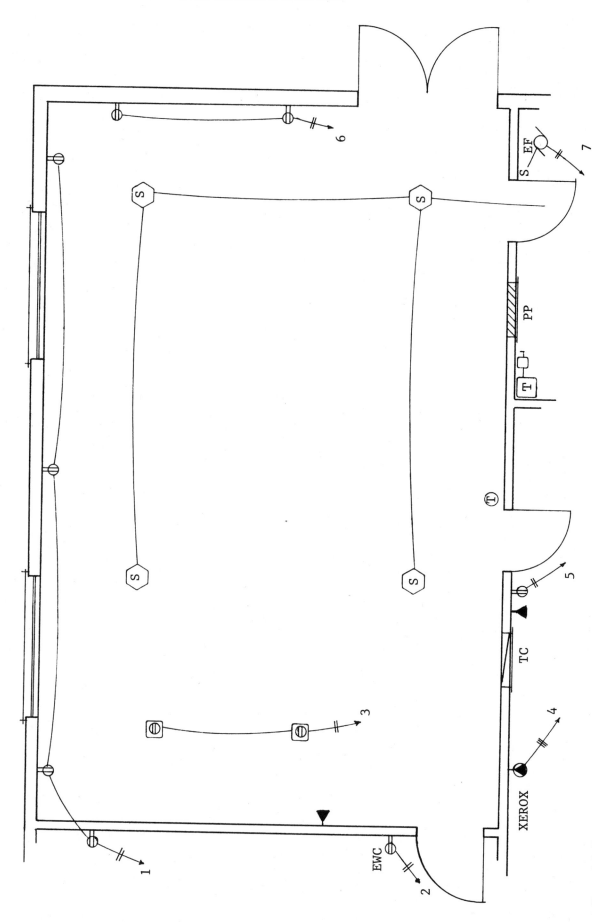

Figure 8-5. Power floor plan shows all equipment and devices requiring electrical connections and includes equipment provided other disciplines. Nonelectrical equipment is described by the discipline providing the item, and location, and electrical service must be coordinated with that discipline.

Receptacles may or may not be circuited, depending upon office practice. When not circuited, the contractor has the option to circuit as necessary in conformance with the requirements of local or national codes. Generally it is considered good engineering practice to fully circuit all devices. In addition to the lighting panel shown on the lighting floor plan (Figure 8-1), two more are shown on the power plan. A panel symbol is used to illustrate the panel and letters are used to identify the individual panel. Actually, the telephone panel is usually called a telephone cabinet. Where more than one panel of a certain type is used, the panels should be identified by a numerical suffix, i. e., PP-1, PP-2, PP-3.

In cases where receptacle locations are important they must be dimensioned from walls, column lines, or other fixed points. Telephone outlets are usually shown on power floor plans except in large, complex installations where a separate telephone floor plan may be developed. Telephone outlets are not circuited as this is performed by the telephone representative. Communication devices are located and described on power plans and, depending upon the project, other electrical and electronic devices may be included.

Figure 8-6 is an electrical power symbol list for the drawing in Figure 8-5. The most important consideration in the development of any symbol list is that dif-

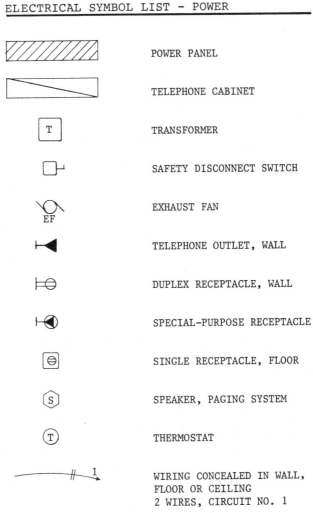

ELECTRICAL SYMBOL LIST - POWER

	POWER PANEL
	TELEPHONE CABINET
T	TRANSFORMER
	SAFETY DISCONNECT SWITCH
EF	EXHAUST FAN
	TELEPHONE OUTLET, WALL
	DUPLEX RECEPTACLE, WALL
	SPECIAL-PURPOSE RECEPTACLE
	SINGLE RECEPTACLE, FLOOR
S	SPEAKER, PAGING SYSTEM
T	THERMOSTAT
1	WIRING CONCEALED IN WALL, FLOOR OR CEILING 2 WIRES, CIRCUIT NO. 1

Figure 8-6. Symbol list for equipment and devices used on power floor plan in Figure 8-5. (Partially from ANSI T32.9-1972)

ferent type symbols be used for different equipment and services. Since there are only so many different shapes that could be used, letter or number inserts may be placed inside the shape to identify the specific symbol. Many A/E firms have developed their own set of symbols that must be used; in some these are from national standards developed by various associations, or could be ones made up by the firm. The symbols should be sized according to the scale of the drawing, as much as practical. For example, lighting fixtures, panels, etc., can be shown to scale; however, small devices such as switches, receptacles, speaker outlets sometimes cannot be drawn to scale since the items may be too small. Also, when letter and number inserts must be used, the symbol must be large enough to permit reasonable sized inserts. Finally, all symbols must be properly identified by appropriate notation. The intent of the use of symbols is drafting shorthand so that drawings may be developed without unnecessary clutter and thus speed drafting work and reduce drafting time. Symbols serve a very useful purpose but to be useful they must be readily identifiable and

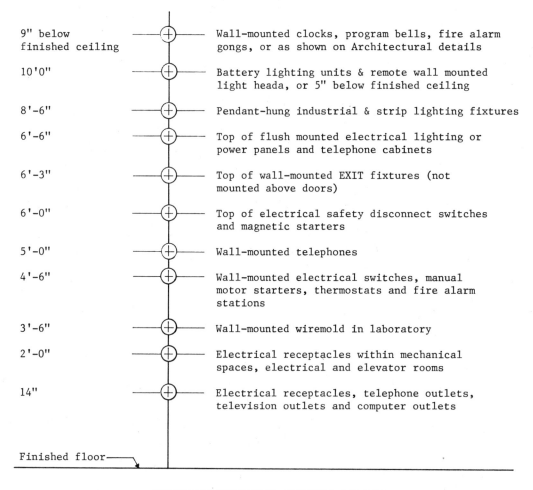

Height	Description
9" below finished ceiling	Wall-mounted clocks, program bells, fire alarm gongs, or as shown on Architectural details
10'0"	Battery lighting units & remote wall mounted light heada, or 5" below finished ceiling
8'-6"	Pendant-hung industrial & strip lighting fixtures
6'-6"	Top of flush mounted electrical lighting or power panels and telephone cabinets
6'-3"	Top of wall-mounted EXIT fixtures (not mounted above doors)
6'-0"	Top of electrical safety disconnect switches and magnetic starters
5'-0"	Wall-mounted telephones
4'-6"	Wall-mounted electrical switches, manual motor starters, thermostats and fire alarm stations
3'-6"	Wall-mounted wiremold in laboratory
2'-0"	Electrical receptacles within mechanical spaces, electrical and elevator rooms
14"	Electrical receptacles, telephone outlets, television outlets and computer outlets

Finished floor

STANDARD MOUNTING HEIGHTS DETAIL

Not to scale

NOTE: Mounting heights to center of
outlets unless otherwise noted

Figure 8-7. A Standard Mounting Heights Detail may be used to indicate elevations of electrical devices on larger, more complex installations. The heights indicated illustrate the methods only, not the actual heights. Some equipment heights are dictated by codes and these requirements must be satisfied.

uniquely different. Drafters must learn the various symbols used by the specific discipline and be able to identify the symbols used by all other disciplines in order to be able to *read* the drawings.

Figure 8-7 provides essential mounting heights and is included, when used, on the same sheet that shows electrical symbols. This information may instead be provided in notes or in the specifications. This information is more important on large complex projects than on drawings for small commercial or residential buildings. When only a few devices require special mounting dimensions, this information is frequently shown directly at that particular device.

3. SINGLE-LINE DIAGRAMS.

Single-line diagrams, also sometimes known as one-line diagrams are an essential part of electrical drawings. A single-line diagram is, as the name implies, a drawing using single lines to illustrate the electrical distribution from the point of power entry to the utilization devices. The drawing may show the incoming service, air break switches, power fuses, transformers, metering, secondary breakers, transformers to utilization voltages, wire and conduit sizes, motor controllers, lighting and power panels, and on to the utilization devices. Single-line diagrams could be complex drawings requiring several illustrations to completely describe the design intent, or may be a small, simple drawing depending upon the size and complexity of the project (Fig. 10-8).

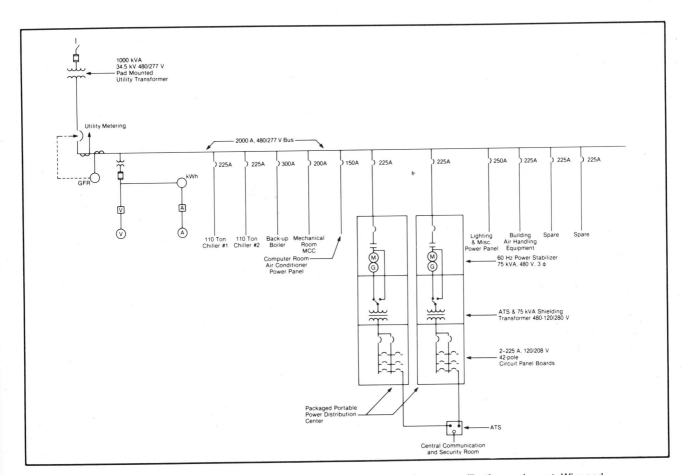

Figure 8-8. Single-line diagram shows distribution from utility transformer to utilization equipment. Wire and conduit sizes are listed on the Electrical Distribution Schedule. (Courtesy of Electrical Consultant Magazine)

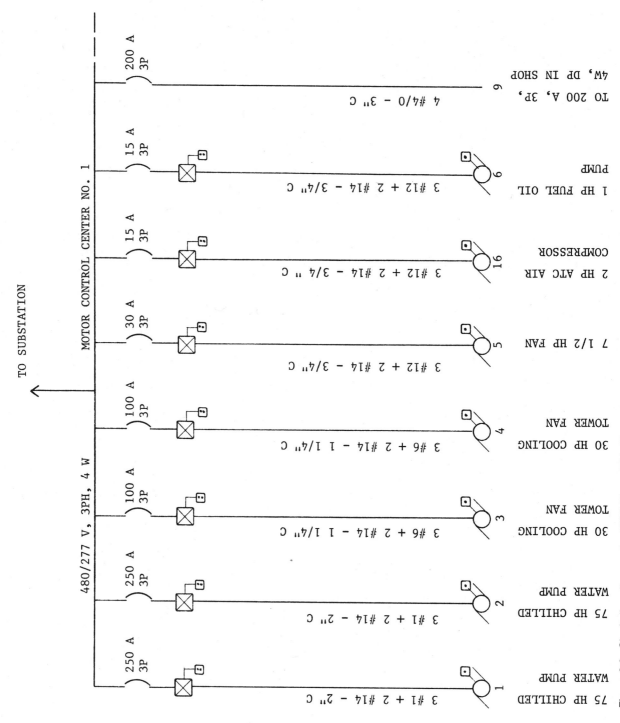

Figure 8-9. Single-line diagram for MCC No. 1. This illustration shows breakers, controllers, push button stations, conduit, and wire sizes. Motors and panels are identified by numbers and this information is noted on MCC No. 1 to indicate locations required.

The single-line diagram is a graphical illustration of the power distribution system. It must be as simple as possible, clearly and concisely drawn, with appropriate symbols to effectively describe the design intent. On larger installations, several drawings are generally necessary. The first describes the incoming service to the first major sub-distribution point. The second drawing may show the distribution from a motor control center, power distribution panel, or utilization transformer (Fig. 8-9).

Essentially, single-line diagrams, properly developed, with one or more drawings as required, will show the entire electrical distribution system from the incoming service to the utilization devices. These drawings do not show physical locations of the equipment; that information is on the associated plan drawings and/or enlarged details. Contractors use plan drawings to determine equipment placement and single-line diagrams for the method of electrical distribution and equipment and device interconnections. The equipment and devices are described in notes on the drawings, in equipment schedules, or in the specifications. Electrical designers and drafters must be fully aware of equipment specified by other disciplines, the location of this equipment, and the required services to the equipment. Additionally, electrical designers and drafters must coordinate electrical plans with single-line diagrams with the electrical section of the specifications.

Single-line diagrams may not be required on small commercial projects and residences because the distribution is usually very simple and can be described by other means. On large, complex projects, single-line diagram are a shorthand technique for describing and illustrating the distribution system. The best method for determining whether or not single-line diagrams are required is to try to describe, by other means, the distribution of power on plan drawings.

Elevations are usually required for motor control centers and substations to show the placement of the various control devices. These are required to illustrate the of the controls for future reference. It is important to number the equipment and the corresponding controls for safety purposes. For example, there should be no question which device controls a specific piece of equipment, i.e., chilled water pump No. 1 is identified on the riser diagram, its control is identified on the controlling device and the equipment is appropriately tagged. Second, the placement of equipment controls should not be left to the arbitrary selection of the contractor. Third, most contractors will demand that this information be included in the contract documents (Fig. 8-10).

All symbols used on electrical drawings must be listed and fully described. Symbol lists must indicate a uniquely identifiable symbol along with a short description. The complete description, in some cases along with a manufacturer's model number is included in the specifications. Sometimes one manufacturer is listed; other times a choice of two or more are included, but in every instance the quality of the particular item is specified (Fig. 8-11).

Single-line diagrams should not be confused with riser diagrams. Riser diagrams are usually developed by the electrical discipline, and show, in single line, the various fire safety and communications devices. There is no way that this information can be adequately shown on plan drawings. All the devices shown on riser diagrams are also shown on plan drawings to indicate their physical locations.

Figure 8-12 is a riser diagram for a multi-floor building and shows the connections of various fire-safety components and how the overall system in inte-

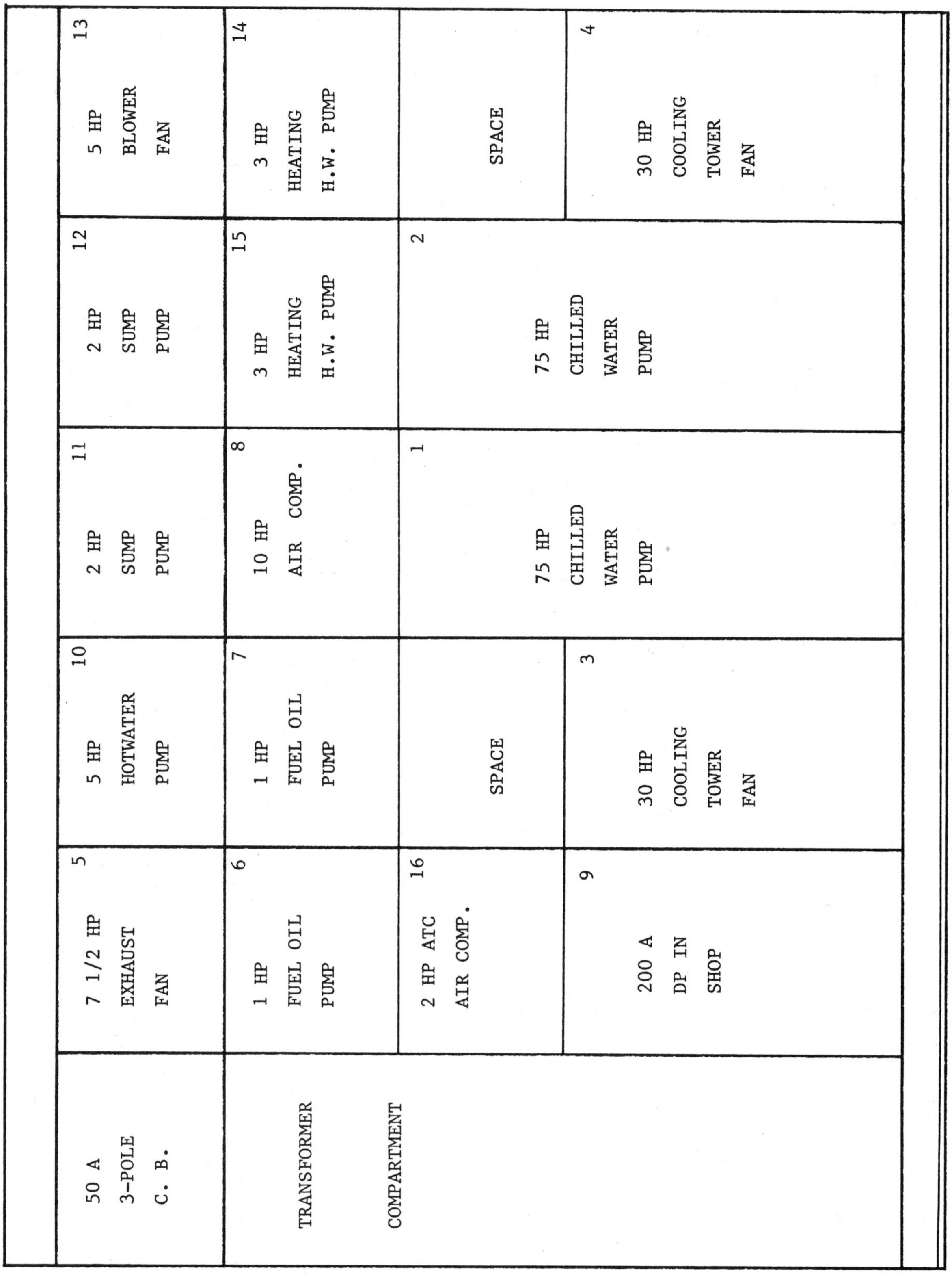

50 A 3-POLE C. B.	5 7 1/2 HP EXHAUST FAN	10 5 HP HOTWATER PUMP	11 2 HP SUMP PUMP	12 2 HP SUMP PUMP	13 5 HP BLOWER FAN
TRANSFORMER COMPARTMENT	6 1 HP FUEL OIL PUMP	7 1 HP FUEL OIL PUMP	8 10 HP AIR COMP.	15 3 HP HEATING H.W. PUMP	14 3 HP HEATING H.W. PUMP
	16 2 HP ATC AIR COMP.	3 SPACE	1 75 HP CHILLED WATER PUMP	2 75 HP CHILLED WATER PUMP	SPACE
	9 200 A DP IN SHOP	30 HP COOLING TOWER FAN	30 HP COOLING TOWER FAN		4 30 HP COOLING TOWER FAN

MOTOR CONTROL CENTER NO. 1

FRONT ELEVATION – NO SCALE

Figure 8-10. Front elevation of MCC No. 1 indicates locations of all control panels.

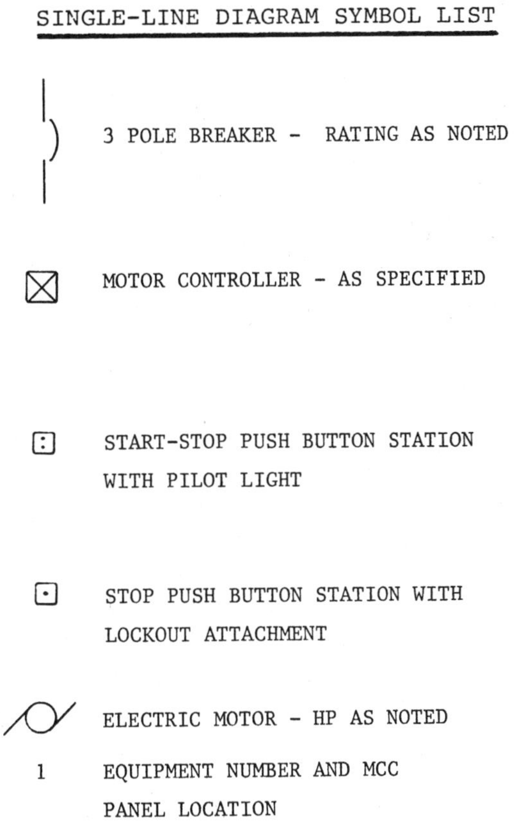

SINGLE-LINE DIAGRAM SYMBOL LIST

3 POLE BREAKER - RATING AS NOTED

MOTOR CONTROLLER - AS SPECIFIED

START-STOP PUSH BUTTON STATION
WITH PILOT LIGHT

STOP PUSH BUTTON STATION WITH
LOCKOUT ATTACHMENT

ELECTRIC MOTOR - HP AS NOTED

1   EQUIPMENT NUMBER AND MCC
PANEL LOCATION

Figure 8-11. Symbols used on single-line diagram in Figure 8-9 are identified and described.

grated. The associated specifications describe the function and operation of the system and plan drawings show device locations (Fig. 8-13).

4. SCHEDULES

Schedules are a simplified means for describing three or more items of similar characteristics in a tabular format. When only one or two types are used these may be more readily described in a short note on the drawing or in the specifications. Most offices use preprinted, adhesive-backed transparent film that is placed on the underside of the drawing. This method has several advantages; first, it permits erasures on the face of the drawing without damaging the form; second, the form provides space for all essential information and all that is necessary is to completely fill in all the spaces; and third, these forms are usually developed from years of experience and permit the completion of the forms by different drafters and designers with the same end result.

Some of the more common schedules used on electrical drawings include:

- Lighting panel
- Lighting fixture
- Power panel
- Panels
- Feeders
- Motor control
- Emergency lighting

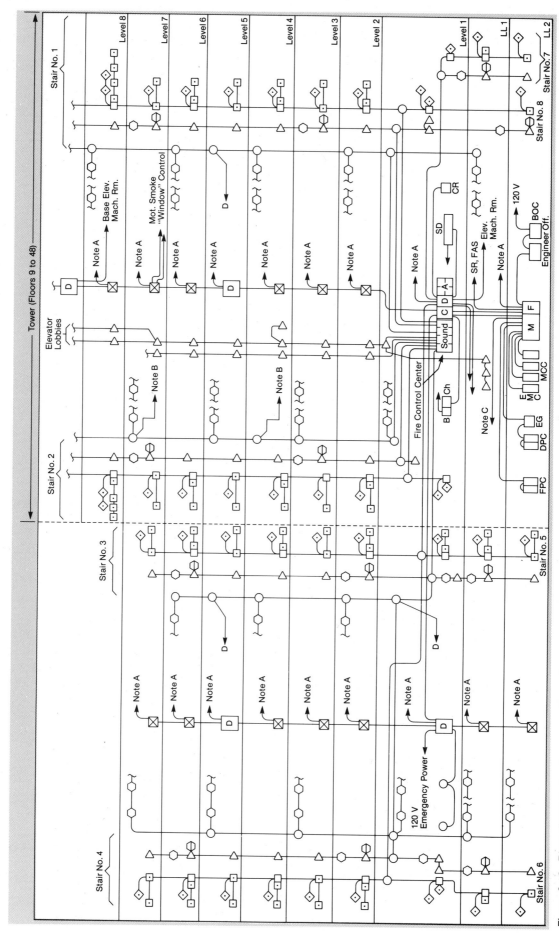

Figure 8-12. Riser diagram showing connections of various fire-safety components and how the overall system is integrated for a multi-floor building. (Courtesy of Electrical Consultant Magazine)

142

Key

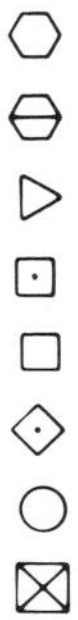

⬡ = Speaker
◐ = Stairwell Emergency Phone
△ = Fireman's Phone Jack
⊡ = Magnetic Door Holder
▢ = Transformer, 277/24V
◇ = Electric Door Lock, 24 Vac
○ = Junction Box
⊠ = Pull Box

Legend

A   = Annunciators
B   = Batteries
BOC = Building Operating Panel
C   = Central Processing Unit
Ch = Battery Chargers
CR = Card Reader
D   = Data Gathering Panel
DPC = Diesel Fire Pump Controller
EG = Emergency Generator
EMC = Emergency Motor Control Center
F   = Fire Alarm Panel
FAS = Fan Air Flow Switch

Legend (Cont.)

FPC = Fire Pump Controller
M   = Mechanical
MCC = Motor Control Center
SD = Security Desk
SR = Smoke Removal

Note A: To manual pull switches, duct smoke detectors, ceiling smoke detectors, sprinkler water flow switches, sprinkler valve supervisory switches, smoke or removal dampers, fan air flow switches, magnetic motor starter relays.

Note B: To elevator half-way boxes for speaker and fireman's phone jack in each car.

Note C: To various metering instruments

Figure 8-13. Symbols and legends for the riser diagram in Figure 8-12. (Courtesy of Electrical Consultant Magazine)

Figure 8-14 is a preprinted lighting fixture schedule that satisfies the purpose of fully describing a wide variety of lighting fixtures. For example, reading from left to right:

Type—Use letter or number for reference to identification with floor plans

Light source—Select the type from the list on the schedule;

> Source type: I—Incandescent
> F—Fluorescent
> MV—Mercury vapor
> MH—Metal halide
> HPS—High pressure Sodium

Mounting*—See note *Contactor to verify type of mounting of all fixtures with room finish schedule.

> Mounting: S—Surface
> R—Recess
> W—Wall
> P—Pendant

Description—Write in the description of the particular type as listed in the manufacturer's catalog

Lamps—Quantity—Write in the number of lamps per fixture.

Type—Write in the type of lamp.

Watts—Write the rated watts of each lamp.

Watts/Fixture—Write in total watts of the fixture.

Volts—Write lamp voltage.

Manufacturer and Catalog No.—Write in selected fixture number.

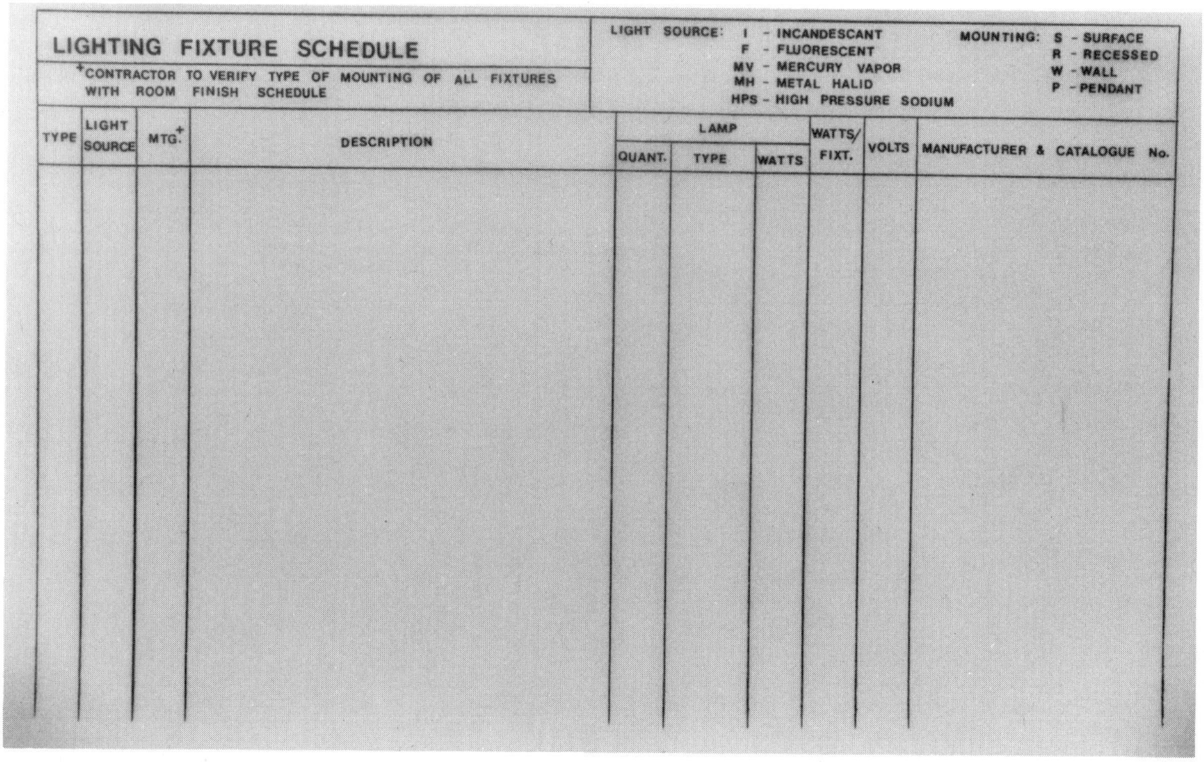

LIGHTING FIXTURE SCHEDULE											LIGHT SOURCE: I - INCANDESCANT F - FLUORESCENT MV - MERCURY VAPOR MH - METAL HALID HPS - HIGH PRESSURE SODIUM	MOUNTING: S - SURFACE R - RECESSED W - WALL P - PENDANT
*CONTRACTOR TO VERIFY TYPE OF MOUNTING OF ALL FIXTURES WITH ROOM FINISH SCHEDULE												
TYPE	LIGHT SOURCE	MTG.*	DESCRIPTION	LAMP			WATTS/ FIXT.	VOLTS	MANUFACTURER & CATALOGUE No.
				QUANT.	TYPE	WATTS			

Figure 8-14. Lighting Fixture Schedule.

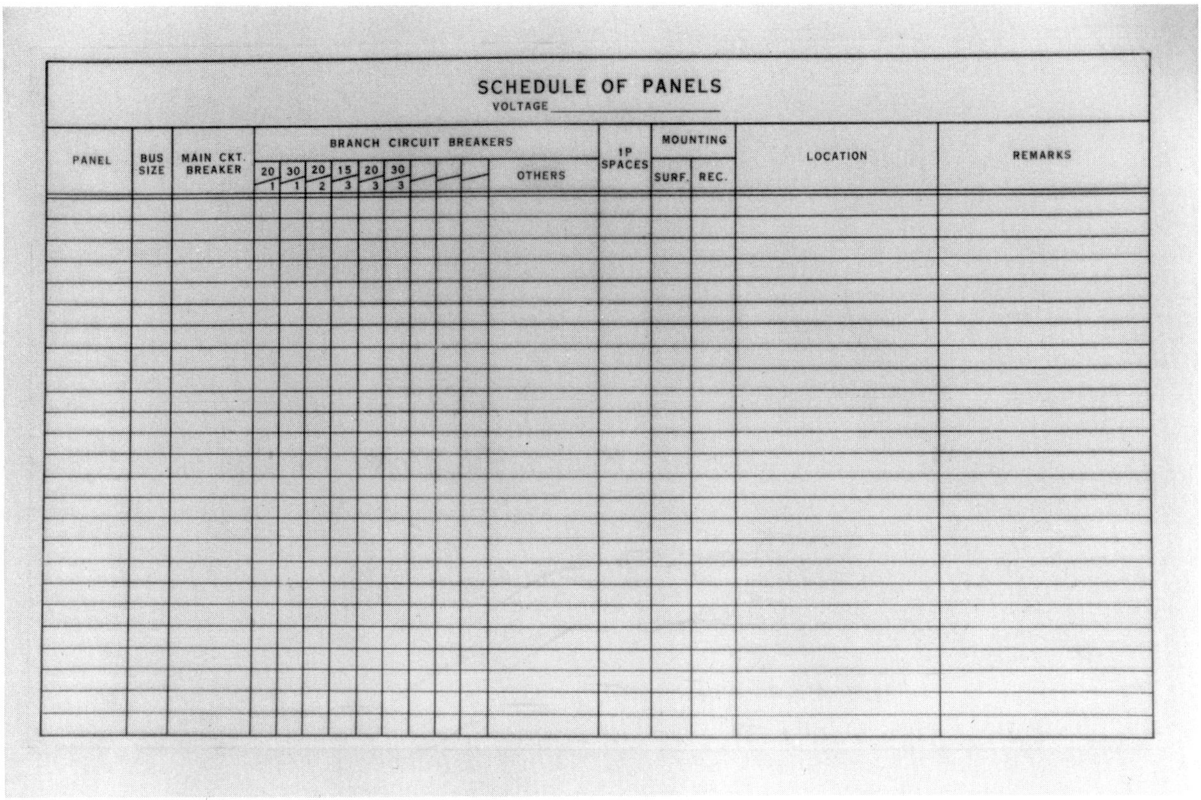

SCHEDULE OF PANELS
VOLTAGE_____

PANEL	BUS SIZE	MAIN CKT. BREAKER	BRANCH CIRCUIT BREAKERS									OTHERS	1P SPACES	MOUNTING		LOCATION	REMARKS
			20 1	30 1	20 2	15 3	20 3	30 3						SURF.	REC.		

Figure 8-15. Schedule of Panels.

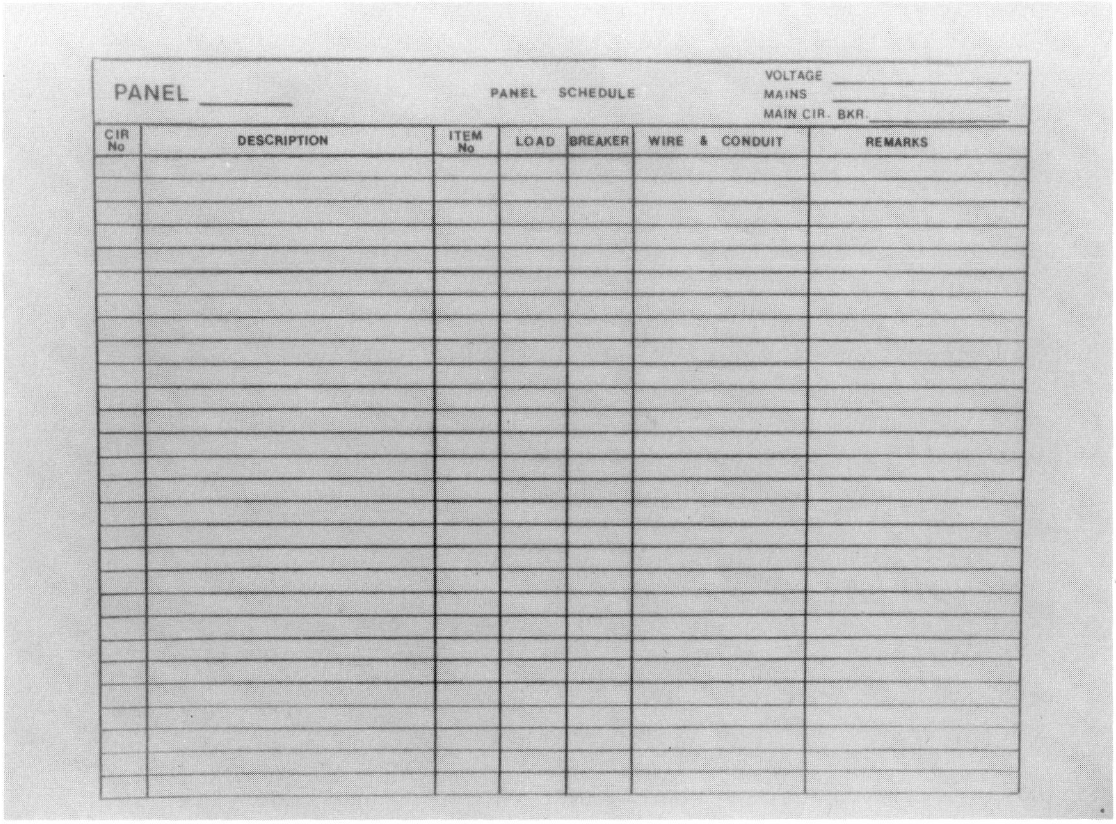

PANEL_____	PANEL SCHEDULE		VOLTAGE _____ MAINS _____ MAIN CIR. BKR. _____				
CIR No	DESCRIPTION	ITEM No	LOAD	BREAKER	WIRE & CONDUIT		REMARKS

Figure 8-16. Panel Schedule.

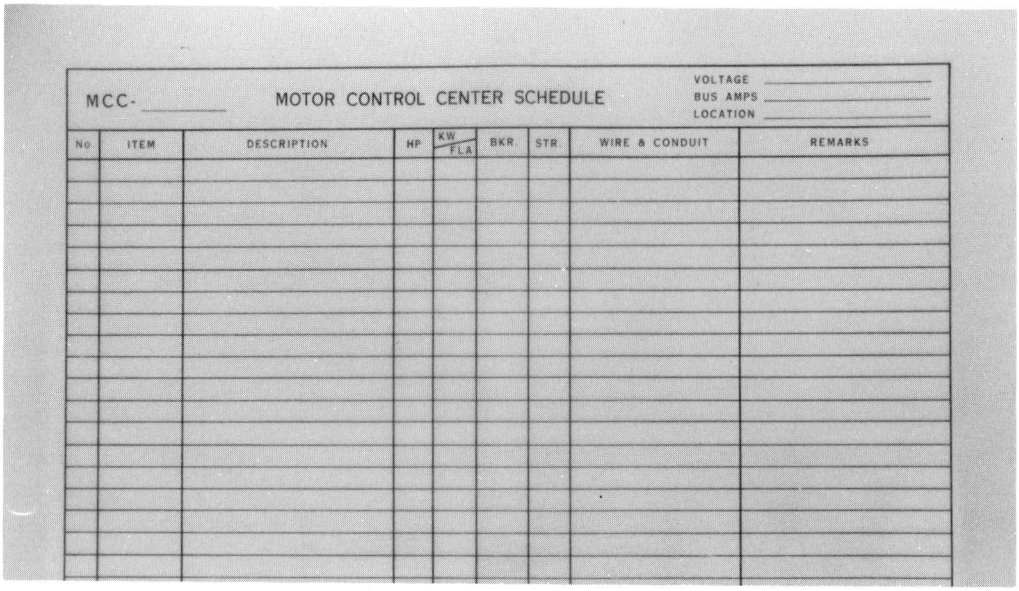

MCC-______ MOTOR CONTROL CENTER SCHEDULE

VOLTAGE ______
BUS AMPS ______
LOCATION ______

No	ITEM	DESCRIPTION	HP	KW / FLA	BKR.	STR.	WIRE & CONDUIT	REMARKS

Figure 8-17. Motor Control Center Schedule.

This form of schedule is particularly well suited for large projects where more than one drafter or designer is assigned to the project. Anyone required to complete this schedule will be able to provide all the essential information in a recognizable format easily identifiable by the contractor, and the possibility of error and omission is greatly reduced.

The Schedule of Panels is used in cases where more than one panel is needed on the project. Here again the form provides columns for insertion of all appropriate information. The left column is used to identify the panels by the same designation used on floor plans, such as, LP-1, LP-2, LP-3, etc. (Fig. 8-15).

The Panel Schedule is used for every panel on the project and to detail the distribution from the panel to the utilization device. The identification on this schedule must match the panel designation on floor plans, and in cases where a Schedule of Panels is used, the designation must match that also. The panel is identified in the upper left-hand corner of the schedule and one line is used for each circuit, and each circuit is appropriately numbered. This tells the contractor what to do as far as circuiting is concerned and also provides a record for future use by the building owner. The contractor is directed to complete a circuit identification card that must be placed in the card holder on the panel (Fig. 8-16).

A Motor Control Center Schedule is used whenever motor control centers are installed. This schedule is seldom found on small commercial projects and is never needed for residential construction (Fig. 8-17).

These schedules provide all the appropriate information deemed essential for the use of the schedules. On small projects, schedules may be drawn directly on the schedule sheet of the set of drawings. Most importantly, schedules are a drafting shorthand technique and could be used to advantage.

Chapter 9
Mechanical (HVAC)

What is shown on mechanical drawings depends upon what is included in the mechanical discipline. On small projects and in small firms, mechanical engineering provides the design of heating, ventilating and air conditioning, plumbing and fire protection. Automatic temperature control (automation or ATC) is usually part of HVAC. In larger firms, mechanical engineering addresses power plant and processing plant design, and all other mechanical engineering is handled separately under HVAC, automation, plumbing and fire protection.

For the purposes of this book, mechanical engineering will be limited to HVAC and associated automation. Plumbing and fire protection are addressed later. There are two reasons for this: first, power plant and processing work are somewhat unique and require special drafting techniques, and second, HVAC is by far more commonly encountered in commercial, institutional, and industrial work. As such it deserves appropriate attention.

Mechanical (HVAC) drawings show all mechanical equipment, duct work, piping and associated valves and pumps, and automatic temperature control. The equipment could include chillers, condensers, cooling towers, packaged and roof-top air conditioners, air handlers, unit heaters, exhaust fans. This equipment is interconnected with pipes or ducts and contains devices such as heating and cooling coils, reheat coils, humidifiers, filters, mixing boxes, variable-air-volume devices, diffusers, registers and grilles.

Schedules, symbols, and abbreviations have an important role in mechanical drafting. Mechanical drawings usually consist of floor and roof plans, enlarged drawings for mechanical rooms, pipe and duct chases, chiller rooms, and other mechanical spaces. On large projects, many typical details are included. In fact sometimes several drawings are devoted to typical details. Some firms have standard details with all the necessary information included on them that is merely copied by the drafters. Other firms use transparent, adhesive-backed, predrawn sheets that are pasted onto the drawing. In this case, the drafter merely selects the appropriate detail from the file and places it on the drawing. Title and reference identification is frequently all that is necessary.

Whenever more than two items have similar characteristics, these are identified and described in schedules. For example, when there are three or more exhaust fans on the project, complete information is detailed on the schedule and the identification, such as, EF-1, EF-2, EF-3, etc., is placed adjacent to the equipment on the drawing. This helps to eliminate extensive notation at each location and simplifies drafting. Sometimes the fan schedule is completed by a typist to reduce drafting time further and to provide a more legible schedule.

The intent of all schedules, symbols, and abbreviations is to simplify drafting and to reduce drafting time.

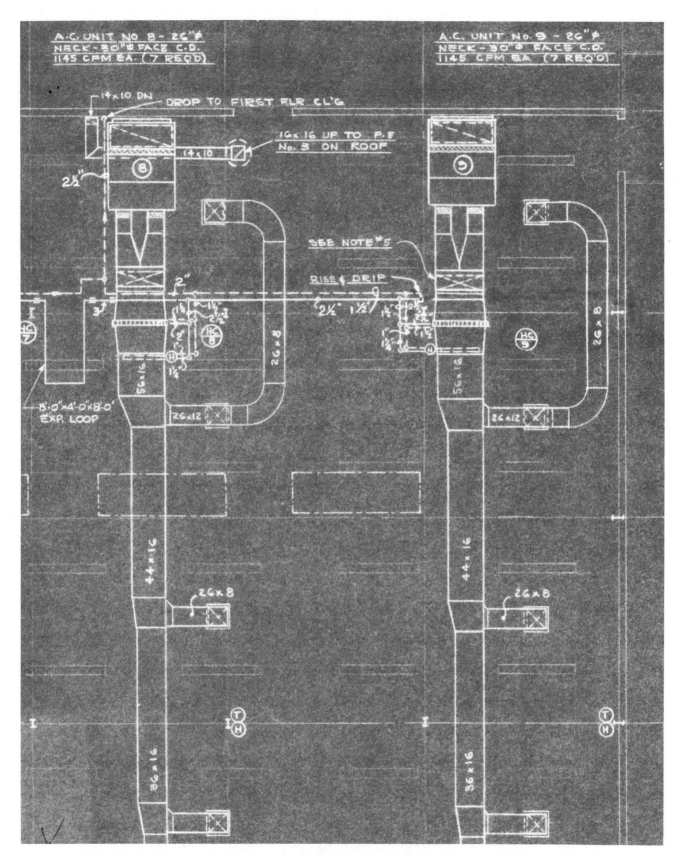

Figure 9-1. HVAC floor plan shows two air conditioning units with double line duct distribution system, including steam heating and reheating coils and humidifiers. Heating coils (HC-8 and HC-9), reheat coils (RH-8 and RH-9), AC-8, AC-9, and humidifiers are described in equipment schedules. (Continued next page)

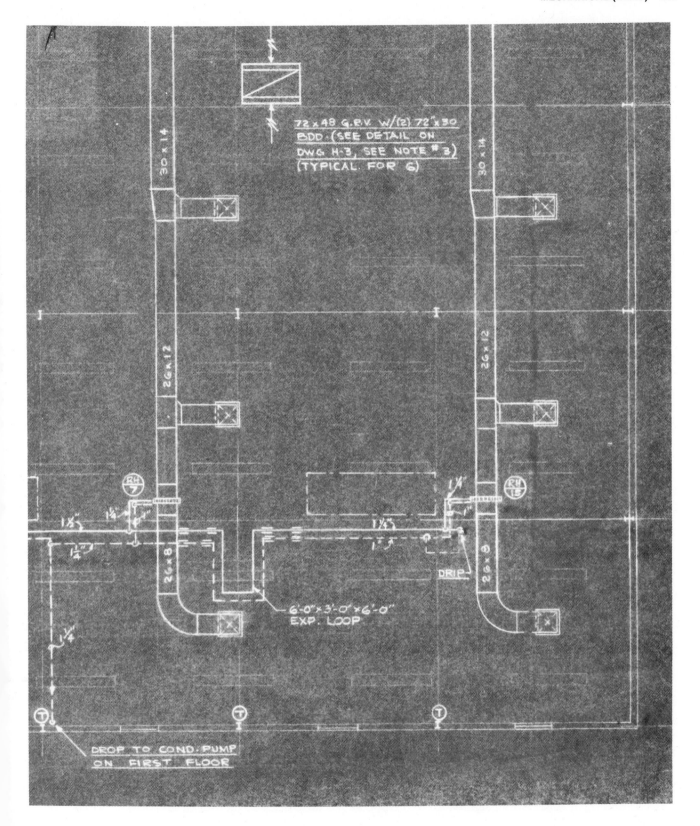

Figure 9-1. (*Continued*)

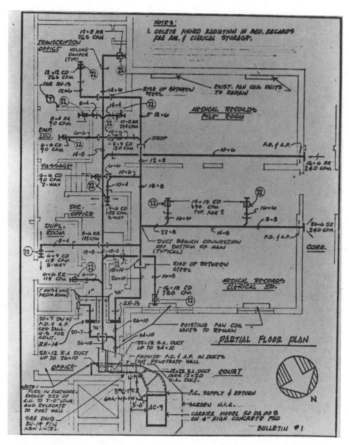

Figure 9-2. This drawing was issued as a Bulletin for a project. It was made on an $8\frac{1}{2} \times 11$ inch sheet and shows the effective use of the single-line duct system. All appropriate components are illustrated and identified.

1. FLOOR PLANS

HVAC floor plans show equipment and associated piping, valves, pumps, ducts, diffusers and registers. Piping is usually shown in single line and pumps and valves are indicated by symbols. Duct work is mostly shown with two lines, drawn to scale, with the actual duct dimensions printed on the duct, for example, 20 × 12. The first figure represents the size of the duct face shown and the second for the side not shown. On a plan view, this means that the duct is 20 inches wide and 12 inches deep. This same duct shown in elevation would be identified as 12 × 20 (Fig. 9-1, Fig. 9-2).

While the electrical drafter is not much concerned about the space between the ceiling and the underside of the structure above, the mechanical drafter must be aware of this space and how the various pipes and ducts fit into the space. In most instances, this space is made as small as practical in order to limit the overall height of the building. Consider, for example, if an additional 12 inches of space is allowed on each floor of a 12-story building. The total height of the building would be increased by 12 feet. This would significantly increase the total cost of the building as 12 additional feet of exterior wall would be required, interior stairs are increased by the same amount, chases and fire walls are similarly increased, and so on. No building owner wants to pay for this increased cost, and no astute A/E firm wastes plenum space unnecessarily.

Special care must be exercised in the design of HVAC air distribution systems and its coordination with other physical equipment in the plenum. The placement of horizontal runs from roof drains, plumbing drains, fire protection piping, electrical lighting fixtures, and other equipment that occupies space in the plenum.

Simple square, rectangular, or circular shapes with appropriate symbols are used to illustrate larger equipment such as chillers, boilers, heat exchangers, air handlers, fan coil units, etc. Large equipment is drawn to scale to indicate the actual space occupied. Also, equipment such as boilers, chillers, heat exchangers that have tubes, etc., that may have to be pulled or removed must show, usually in dashed line form, the space reserved for the function. Sometimes this equipment is dimensioned, but not as a normal practice. Small equipment such as valves, pumps, fans, and control devices, are drawn large enough

HVAC PIPE LINE SYMBOLS AND DESCRIPTION

LINE SYMBOLS	DESCRIPTION
———————	HOT WATER HEATING SUPPLY
— — — — —	HOT WATER HEATING RETURN
——— CHWS ———	CHILLED WATER SUPPLY
— — CHWR — —	CHILLED WATER RETURN
——— CWS ———	CONDENSER WATER SUPPLY
—— — CWR —— —	CONDENSER WATER RETURN
——— FOS ———	FUEL OIL SUPPLY
—— FOR — —	FUEL OIL RETURN
—— — FOV — —	FUEL OIL VENT
——— ———	ANY SUPPLY LINE WITH APPROPRIATE DESIGNATION
— — — —	ANY RETURN LINE WITH APPROPRIATE DESIGNATION

Figure 9-3. Some of the more common pipe line symbols and letter designations used on HVAC drawings. (From ASA Z32.2.3-1949)

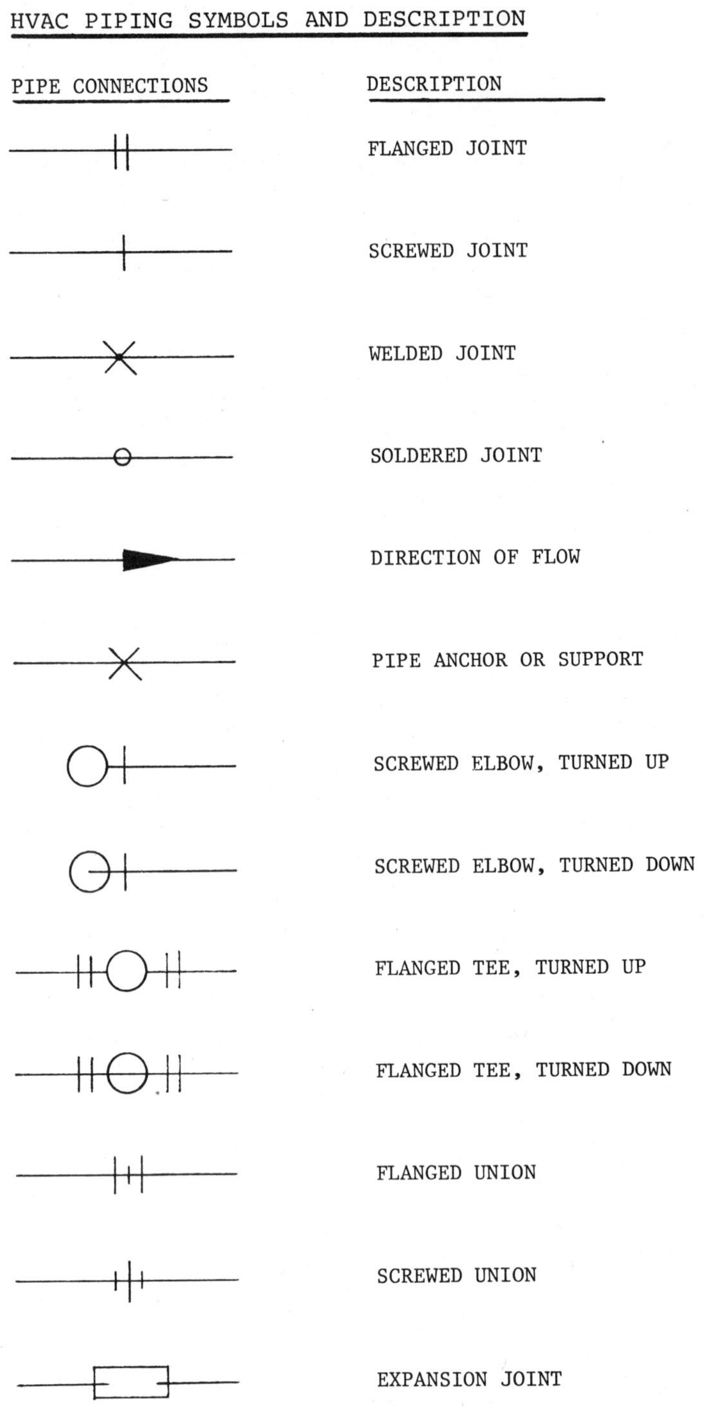

HVAC PIPING SYMBOLS AND DESCRIPTION

PIPE CONNECTIONS	DESCRIPTION
——————╫——————	FLANGED JOINT
——————┼——————	SCREWED JOINT
——————✕——————	WELDED JOINT
——————○——————	SOLDERED JOINT
——————▶——————	DIRECTION OF FLOW
——————✕——————	PIPE ANCHOR OR SUPPORT
○┤——————	SCREWED ELBOW, TURNED UP
⊖┤——————	SCREWED ELBOW, TURNED DOWN
——╫○╫——	FLANGED TEE, TURNED UP
——╫○╫——	FLANGED TEE, TURNED DOWN
——┤├┤——	FLANGED UNION
——┤├┤——	SCREWED UNION
——┤▭├——	EXPANSION JOINT

Figure 9-4. Symbols are used to indicate the various methods of piping connections at tees, elbows, and other fittings, and for other means of installation. (From ASA Z32.2.3-1949 and ASA Z32.2.4-1949)

to be readily recognized, and many times the symbol is larger than the scale for the device. For example, a small valve, say 3 inches in size, cannot be drawn to scale at ⅛"—1"—0". Such devices must be drawn reasonably larger so that an appropriate sized symbol can be used. Also, small diameter pipe is indicated by lines that are usually larger than scale size. This practice does not create confusion because pipe line sizes are indicated on drawings at regular intervals.

Mechanical drawings differ from electrical in the "home run" techniques are not used. All ducts and pipes are shown in their entirety. For example, pipe lines from a chiller to an air handler and return are drawn from the chiller to the air handler and back. On HVAC plan drawings, even large-sized pipes are shown in single line. However, on enlarged detail drawings, these large pipes may be drawn in double line where necessary to indicate the space occupied by the pipes. When a single heavy line is used to indicate pipe runs, letter designations are placed in series with the lines, in addition to the standard pipe line symbols for positive identification (Fig. 9-3, Fig. 9-4).

Solid lines are usually used for all supply pipes and dashed lines for returns, both with appropriate letter designations indicated, for example, ——— CHWS ——— for chilled water supply, and ———— CHWR ———— for chilled water return.

Most A/E firms have established a complete list of pipe line symbols that are standard for the office. Some develop their own, while others use various national or association standards. In cases where pipe line symbols are not available for the type of line intended, letter designations may be developed to identify the line and this symbol must be included in the pipe line symbol list.

Engineering drawings are graphical presentations, and frequently pipe lines

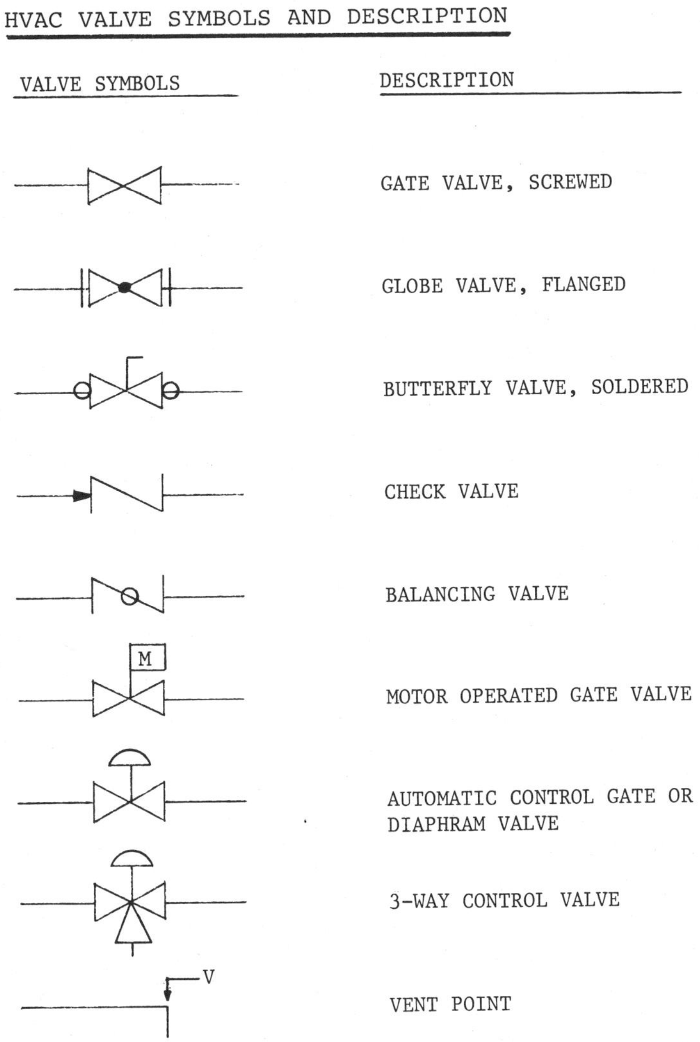

HVAC VALVE SYMBOLS AND DESCRIPTION

VALVE SYMBOLS	DESCRIPTION
	GATE VALVE, SCREWED
	GLOBE VALVE, FLANGED
	BUTTERFLY VALVE, SOLDERED
	CHECK VALVE
	BALANCING VALVE
	MOTOR OPERATED GATE VALVE
	AUTOMATIC CONTROL GATE OR DIAPHRAM VALVE
	3-WAY CONTROL VALVE
	VENT POINT

Figure 9-5. Symbols are used for the myriad of values used in piping systems. Some of the more common types are included here. Any symbol used on drawings must be listed and described on the valve symbol list. (From ASA Z32.2.3-1949)

HVAC AIR DISTRIBUTION SYMBOLS AND DESCRIPTIONS

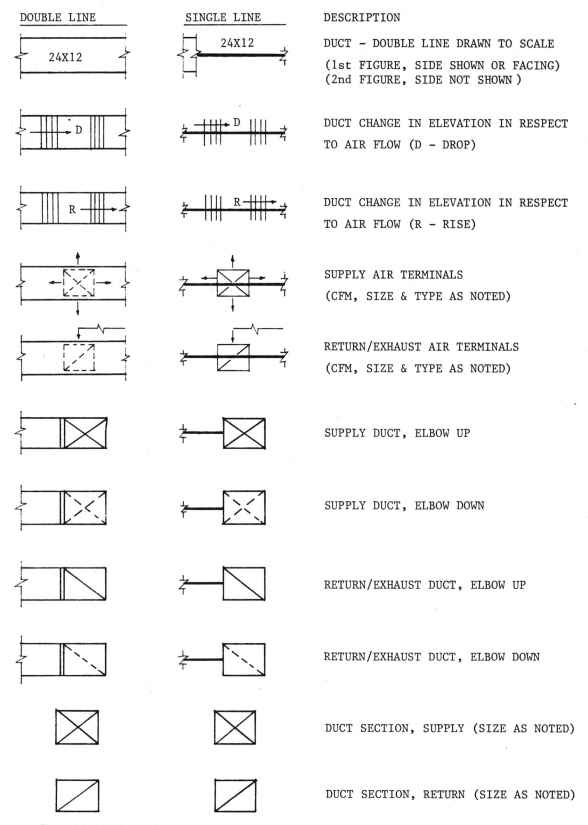

Figure 9-6. HVAC air distribution duct system may be drawn either double line or single line, depending upon circumstances and/or office practice. Duct sizes are always indicated and terminals are shown. (Partially from ASA Z32.2.4-1949)

are spread apart a greater distance or are moved further from walls than they will actually be installed. This may be required to illustrate the intent of the design properly and to provide space for the letter designation.

Symbols have been developed for all common valves, and these are inserted into the proper location in the pipe runs. Special symbols are used at the terminals of valves to indicate the type of joint or connection. A single vertical line is placed across the pipe line to indicate a screwed connection at tees, elbows, etc., but this vertical line is not used at the valve; here the connection is implied. Two vertical lines are used for flanged connections at tees, elbows, etc., but only one is used at the various valves since the valve contains the mating flange. (Fig. 9-5).

Air distribution duct work on plan drawings may be shown in either double line or single line, depending upon office practice or preference. Duct work on enlarged detail drawings is always shown in double line, and all double line ducts are drawn to scale. Additionally, the duct is dimensioned to indicate the size of the duct with the first figure indicating the side shown and the second figure indicating the side not shown. This is used with both double-line and single-line methods. While the single line may be used to indicate the duct run in the distribution space, a double line is used in and around the equipment and to some point beyond. At that point the double line is appropriately terminated

HVAC AIR DISTRIBUTION SYMBOLS AND DESCRIPTIONS

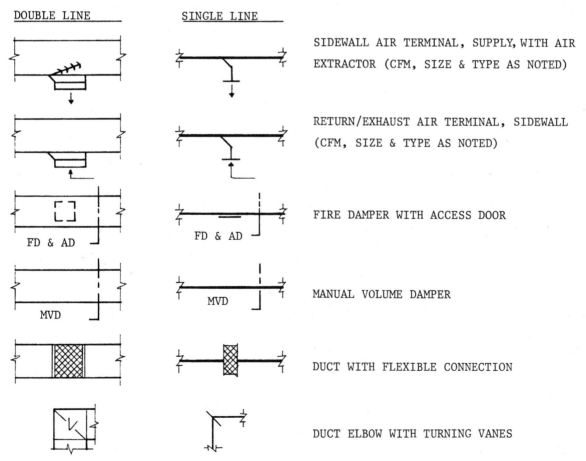

Figure 9-7. Every HVAC air distribution symbol used on drawings, including terminals, direction of air flow, dampers, access doors, flexible connections, turning vanes, etc., must be included on the symbol list. (Partially from ASA Z32.2.4-1949)

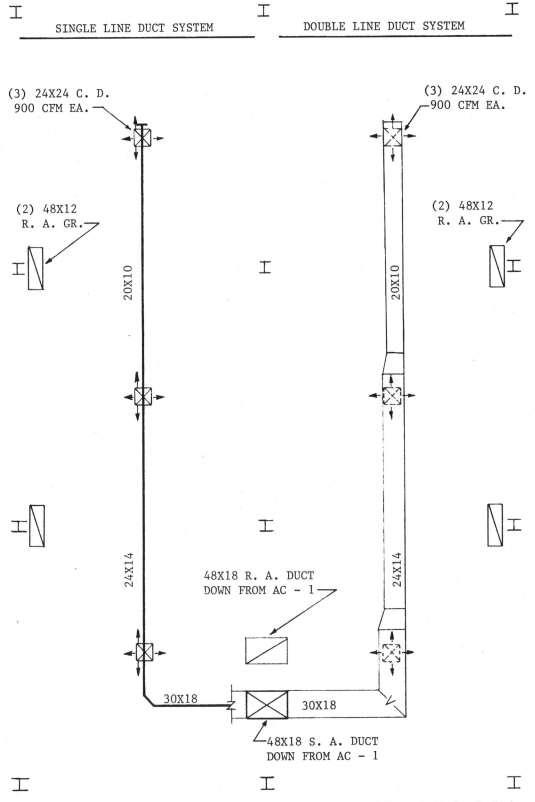

Figure 9-8. Example of a typical HVAC air distribution system. Either the single-line or double-line distribution method may be used effectively as long as all appropriate notes are included.

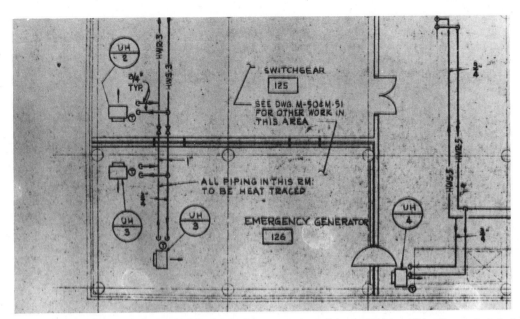

Figure 9-9. This drawing shows a piping distribution system for unit heaters. Note unit heater symbols indicate the direction of air flow. Pipe line sizes are noted and hot water supply and return lines are identified by HWS and HWR. The numbers 3 and 5 identify pipe line branches; this may not be required on some projects. Unit heater data are listed on the unit heater schedule.

and the single line may be used for distribution. Appropriate symbols are used to indicate duct elevation changes, air terminal devices, connection points, take-offs, and other devices and equipment (Fig. 9-6, Fig. 9-7).

Figure 9-8 illustrates the use of both double-line and single-line distribution within a space. Note the method of showing duct sizes and indicating air terminals. Air terminals are further described by size and the air quantity is noted. Appropriate symbols are used for the section cuts of the duct with appropriate notes, i. e., "48 × 18 Return Air Duct Down from AC-1." In this particular case, the air conditioner is on the roof, and the roof plan will show the location and size of the unit. Also see Fig. 9-9 and Fig. 9-10.

2. ENLARGED DETAIL DRAWINGS

Enlarged engineering detail drawings are large-scaled drawings of equipment and spaces used to show the location of the various pieces of equipment and to illustrate the means of assembly in a more detailed manner than can be shown on the small-scale floor plans. This technique is particularly important for spaces such as mechanical rooms, air-handling rooms, chiller rooms, shafts, chases, and other special spaces. These enlarged drawings usually include plan and elevation views and one or more sections required to describe fully the placement of the equipment, run of the duct, and installation of piping, hangers, valves, etc.

On floor plans drawn at ¹⁄₁₆-inch scale and even sometimes on drawings at ¹⁄₈-inch scale, the areas mentioned above are left blank with a note "See Main Fan Room Detail, M-32." This technique is used in cases where the fan room may be crowded and the configuration complex. It would serve no useful purpose and valuable drafting time would be wasted in trying to make a reasonable representation of the space use, particularly when an enlarged detail is planned to be included. In some cases on ¹⁄₈-inch scale floor plans and in all cases on ¹⁄₄-

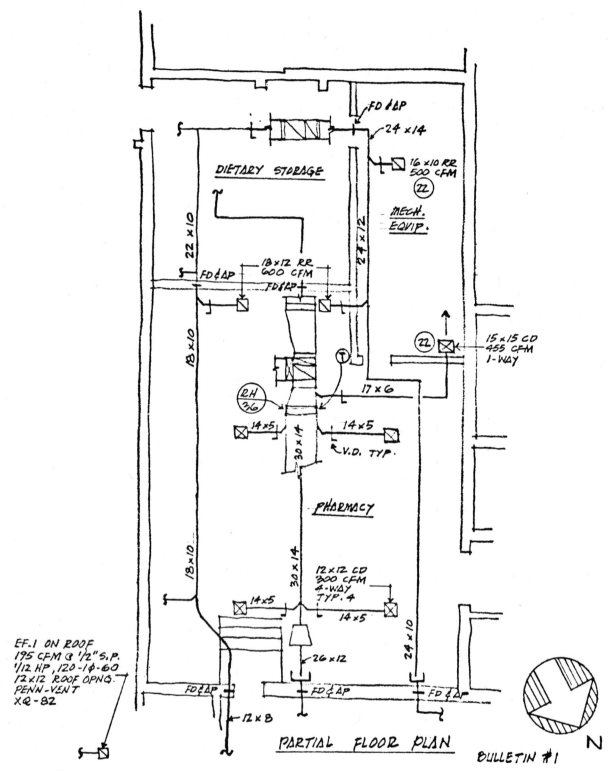

Figure 9-10. Typical freehand sketch drawn by a project engineer is used by drafters to develop hard-line drawings.

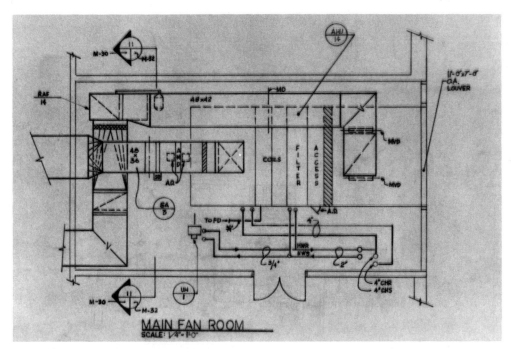

Figure 9-11. This enlarged plan view detail drawing of a fan room was made at a $\frac{1}{4}$-inch scale. All duct work and piping is shown as viewed from above. One section cut is made at the location indicated by the cut line, and the section view is in the direction of the arrow.

inch scale floor plans, the plan view may be completely drawn. Then enlarged drawings of the plan view is unnecessary and only elevation and sections need to be drawn. In most instances in HVAC work, ¼-inch scale enlarged drawings will suffice. Only in very special cases are larger scales used.

Figure 9-11 is a good example of separate detail drawings on the detail sheet. On this project the Main Fan Room space was left blank except for the duct leaving the space and the location of the pipe sleeves and a note "See Main Fan Room Detail, M032." On a separate sheet the ¼-inch plan drawing was developed and this is sufficient to illustrate the required plan view information. Since the duct leaves the space at only one end, only one section cut was needed. The location of the cut and the direction of the view are indicated by the cut line and associated arrow. The circle symbol indicates that "Section 11" (upper number) of the Main Fan Room on Drawing M-30 is shown on Drawing M-32.

Figure 9-12 is the section cut indicated on Figure 9-11. In order to save drafting time a scale is not shown on this section, because ALL details on Drawing M-32 are at the same scale, and this scale is noted in the title box. In other cases where different scales are used on any one sheet, each detail or section must contain the scale used. These two drawings adequately illustrate the design and installation intent.

Figures 9-13 and 9-14 are two section cuts through another fan room on the project. In this instance, the designer determined that two sections were required, in addition to the floor plan, to adequately illustrate the equipment placement and the duct runs.

One essential consideration in drafting is to not waste drafting time by making unnecessary drawings and details. Enlarged drawings should be made only

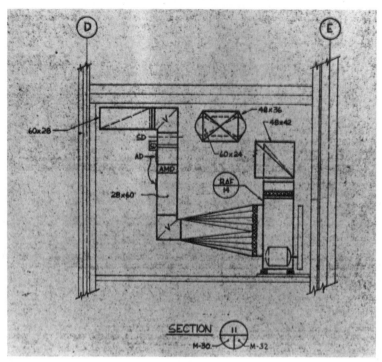

Figure 9-12. This drawing is the section cut indicated on Figure 9-11. Note the correct use of HVAC duct symbols.

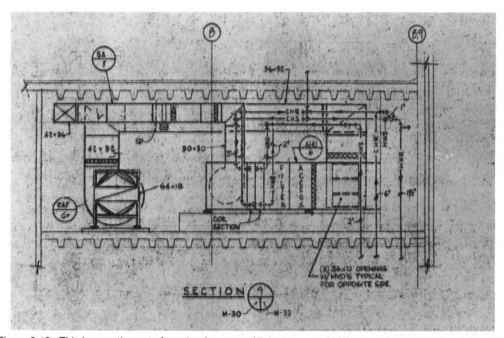

Figure 9-13. This is a section cut of another fan room. All duct runs and piping are shown, and the air handler unit and return fan are correctly identified. Duct sizes indicated by leaders is read to mean that the first number is the size of the duct touched by the arrow and the second number is the size of the perpendicular side of the duct.

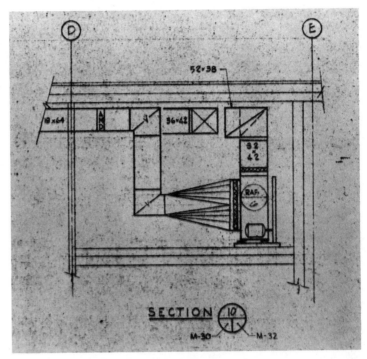

Figure 9-14. This drawing is another section cut of the same fan room where the first cut is shown in Figure 9-13. These two cuts plus the enlarged plan view of the fan room fully describe the equipment placement and duct and pipe runs.

when necessary; never make enlarged drawings merely because it seems like a good idea. Please note that the size of the duct leaving the fan room on Figure 9-11 is not shown. This information is included on Section 11 and was also noted on the floor plan where the duct penetrated the wall. Unnecessary drafting wastes time and could lead to error.

3. TYPICAL DETAILS

Typical details are exactly what the name implies; they are typical drawings of equipment, devices, systems, piping connections, installation methods, etc. included on the drawings for illustrative purposes. The drawings are not made to any scale, since they are typical, enlarged drawings. *Typical detail drawings* are different from *enlarged detail drawings:* the first is a typical drawing showing means and methods of installation, while the second is an exact scaled drawing of some equipment or space.

Typical details are an essential part of mechanical (HVAC) drafting and design. First, the small scale of floor plans usually does not permit the drawing of the necessary intricate details at the equipment or device to describe the method of connections fully. Second, why waste time repeating the detail for every similar piece of equipment when one detail on the typical detail sheet will suffice. Take, for example, the case of a unit heater. Some may be horizontal, some vertical, some water heated and some steam heated. The floor plan usually contains a typical symbol for the unit heater (a box with the designation UH-1) with supply and return pipe lines to the unit. Hot water heated units require a gate valve, a balancing valve, two unions at the unit, drain line with valve and pipe cap, and may require an air vent (Fig. 9-16). Steam-heated units require gate valves, unions, a steam trap, strainer and drain leg (Fig. 9-17). On large projects, there may be many unit heaters, and one typical detail will serve the purpose of describing the detail for every one. Many A/E firms will have the typical detail drawn once and reproduced on adhesive-backed transparencies, which are merely affixed to the appropriate detail sheet. Every detail on every project is identical and the time required to place this detail on the drawing is reduced to a few minutes. Hours of drafting time are eliminated. The following is a list of typical HVAC details that may be found on a large building project:

HVAC TYPICAL DETAILS

a. Foundation pad
b. Inertia pad
c. Fuel oil storage tank installation
d. Low pressure steam trap assembly
e. High pressure steam trap assembly
f. Single steam pressure regulating station
g. Dual steam pressure regulating station
h. Steam control valve assembly
i. Hot water control valve assembly
j. Condensate pump installation
k. Duplex sump type condensate pump
l. Booster and inline pump piping
m. Hot water converter piping diagram
n. Double suction pump piping
o. End suction pump piping
p. Convector piping
q. Steam heating coil piping
r. Chilled water cooling coil piping
s. Hot water heating coil piping
t. Fan coil unit piping
u. Unit heater piping
v. Compression tank
w. Open expansion tank
x. Fire damper mounting (in duct) (Fig. 9-19)
y. Humidifier manifold piping (in duct)
z. Heating coil mounting (in duct) (Fig. 9-18)

Figure 9-15 is a partial floor plan showing unit heaters and the associated piping. The unit heaters here are indicated by a simple symbol, with the supply and return piping terminating at the unit; no complex piping connection details are shown at the unit. The piping connections are detailed only ONCE in a typical detail as shown on Figure 9-16. To simplify drafting on this project further, the boiler room detail was not shown on this small-scale floor plan. Since it was anticipated that an enlarged detail drawing plan view would be required, drafting time was not wasted by showing the boiler room twice, in two places, at two different scales. The boiler room detail was handled simply by the note "See Dwg M-50 & M-51 For Other Work in This Area."

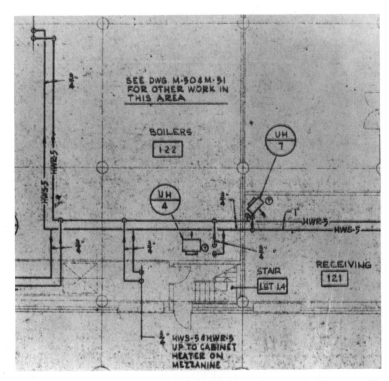

Figure 9-15. Typical partial floor plan showing piping runs terminating at the unit heaters. Pipe size is noted, and the piping system is identified in accordance with appropriate pipe line symbols. The connections at the fan are not shown on this small-scale drawing. The means of connections are fully illustrated on the typical details for Hot Water Heated Unit Heater.

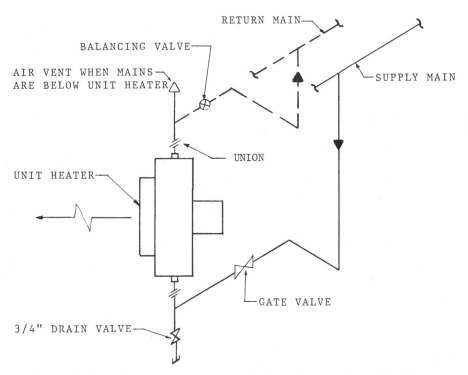

Figure 9-16. Typical piping details for a propeller-driven Hot water Heated Unit Heater.

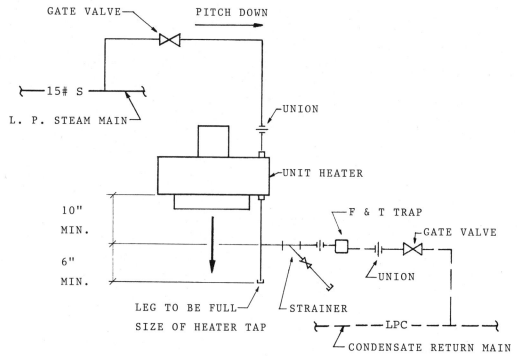

Figure 9-17. Typical piping details for a propeller-driven Steam-Heated Unit Heater.

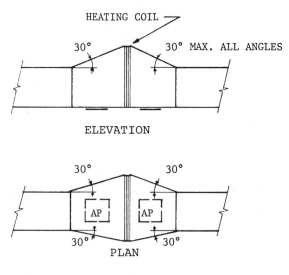

Figure 9-18. Typical mounting details for a heating coil in duct work. Floor plans show the location of the heating coils and this detail describes installation methods.

4. SCHEDULES

Equipment schedules are used far more extensively on mechanical (HVAC) drawings than on any other type of engineering drawings and probably as much as on architectural drawings. The use of schedules provides a convenient and simple method for describing equipment. Detailed equipment information must

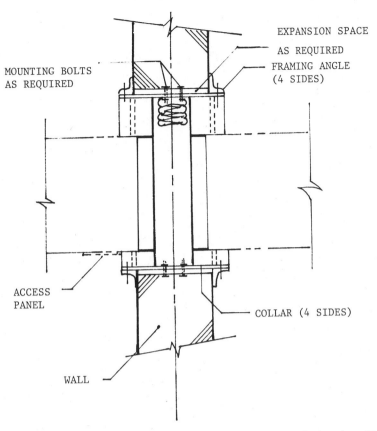

Figure 9-19. Typical installation details for fire dampers. Floor plans show the location of fire dampers.

be provided somewhere in the contract documents. In some offices, this is written into the specifications; in others, it is included on the drawings. Some large, more complex equipment such as transformers, substations, chillers, cooling towers, etc., that require extensive details, are described in the specifications. In many cases, the electrical and mechanical characteristics of this equipment are also included on the drawings. The myriad of other equipment is usually best described on the drawings.

This can be done in any of several different ways. The information can be noted at the location of the equipment, as shown on Figure 9-20. This is convenient for the drafter and designer and handy for the contractor, but on many drawings this creates unnecessary clutter. A much better method is to indicate the information by note numbering, with leaders to the point of application and with the actual note placed outside of the drawing body as shown in Figure 9-21. In this case, the noted information may be typed on adhesive-backed transparent sheets and applied to the drawing in the space reserved for notes, usually along the right-hand edge of the sheet. However, in either case, where to place the total equipment information and the sequence of the information are usually at the discretion of the drafter or designer.

Some brief information, particularly that which is instructional for the contractor, such as "L. P. Trap Ass'y" and "Drop line below 2nd Fl" are frequently best noted at the point of application as shown in Figure 9-20. Longer, more detailed instructions should be included in the list of notes outside the drawing proper, as shown in Figure 9-21.

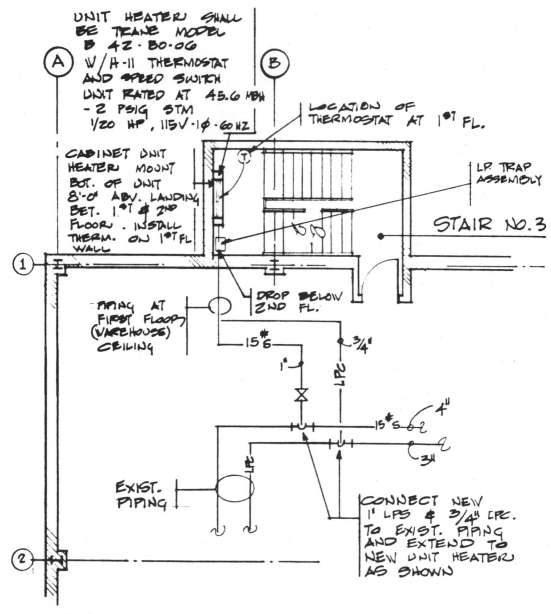

Figure 9-20. Partial floor plan shows the addition of a unit heater, associated piping, and thermostat. All appropriate information is included at the point of application.

Equipment descriptions and characteristics are most frequently listed in equipment schedules, particularly when there are several similar items used in the building project. A properly developed equipment schedule provides adequate space for all essential information, all the drafters and designers have to do to fill in all the spaces (see Figure 9-22). Second, this information is listed in an orderly sequence so that the contractor can quickly and easily locate the appropriate characteristics of the equipment. Further, with an equipment schedule all that is required at the physical location of the equipment is the symbol, UH-1, UH-2, etc. Finally, many large firms use adhesive-backed transparent sheets on which the information may be typed to provide legible copy; this sheet is then affixed to the drawing. Or the transparencies may be affixed to the drawing and the information hand lettered by drafters or designers.

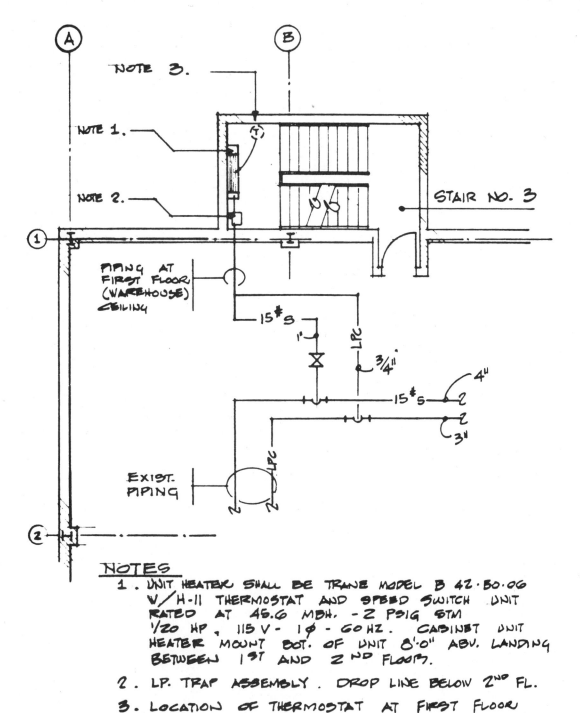

NOTE 3.

NOTE 1.

NOTE 2.

STAIR NO. 3

PIPING AT
FIRST FLOOR
(WAREHOUSE)
CEILING

15#S

1"

LPC

3/4"

15#S

4"

2

2

3"

EXIST.
PIPING

LPC

NOTES
1. UNIT HEATER SHALL BE TRANE MODEL B 42·50·06
   V/H-11 THERMOSTAT AND SPEED SWITCH UNIT
   RATED AT 45.6 MBH. -2 PSIG STM
   1/20 HP, 115V - 1∅ - 60HZ. CABINET UNIT
   HEATER MOUNT BOT. OF UNIT 8'·0" ABV. LANDING
   BETWEEN 1ST AND 2ND FLOORS.

2. LP. TRAP ASSEMBLY. DROP LINE BELOW 2ND FL.

3. LOCATION OF THERMOSTAT AT FIRST FLOOR

Figure 9-21. Same partial floor plan shown in Figure 9-20, but in this case all appropriate information is provided under NOTES away from the point of application. These two drawings illustrate two different methods for describing equipment and providing instructions to the contractor.

Firms that do not use adhesive-backed sheets develop equipment schedules on the appropriate drawing and the necessary information is inserted into the proper space. These schedules must be adequately sized to provide space for the information. The title of the schedule should be larger and more prominent than the lettering used in the body of the schedule. Usually, ⅛-inch lettering is adequate for the general information; the title should be about twice as large,

UNIT HEATER SCHEDULE						
SYMBOL	LOCATION	TYPE	FAN CAPACITY			
			CFM &70°F	HP	ELECTRICAL SERVICE	RPM
UH-1	STORAGE	VERTICAL	1200	1/15	115-1-60	1550
UH-2	EQ.ROOM	HORIZONTAL	815	1/20	115-1-60	1550

COIL CAPACITY				REMARKS	BASIS OF DESIGN
MBH	ENT. AIR °F	STEAM #/HR	STEAM PRESS. PSIG		
93.75	50	100	15	ROOM T-STAT TO CONTROL MOTOR	XYZ CO. AA-225
60.5	50	64	15	"	" . AB-125

Figure 9-22. Typical unit heater schedule. All necessary information is listed in an orderly manner, for example; Unit Heater No. 1 is located in the storage room. It is a vertical type rated at 1200 cfm, 1550 rpm, with a $\frac{1}{15}$ hp motor that requires a 115 volt, single-phase, 60-cycle electrical service. Coil capacity is 93.75 MBH with 50°F air entering the coil that uses 100 lbs of 15 psig steam. The room thermostat is to control the motor, and the basis of design is XYZ Co. AA-225. The HVAC floor plan contains only the circled-up symbol UH-1 with a leader to the unit.

or ¼-inch. With single-line lettering of ⅛-inch size, the horizontal lines should be placed about ¼-inch apart. When two lines of lettering are needed, the horizontal lines should be spaced at about ½-inch to provide enough room for the lettering to avoid crowding. Schedules must contain all pertinent information and the lettering must be legible.

The equipment symbol is placed in the first column on the left of the schedule; this symbol must be identified in the symbol list and is placed adjacent to the specific piece of equipment. The second column indicates the location. On smaller projects, it may be sufficient to note "Storage", but on large projects where several rooms have the same room name, it may be necessary to name the space and also note its identifying number, "Storage Room 105".

With very few exceptions, mechanical equipment uses some form of electrical input. This may be an electrical service to motors or resistance elements. In case of a motor, the horsepower, voltage, phase, and cycles are listed. In case of a resistance element, the horsepower is replaced by wattage.

All the pertinent data of the mechanical portion of the equipment are listed. Take, for example, a unit heater schedule. A unit heater has a fan and a coil, therefore, the electrical information is listed under "Fan Capacity" and the mechanical under "Coil Capacity" (Fig. 9-22).

FAN SCHEDULE										
SYMBOL	SERVICE	LOCATION	CFM	TOTAL PRESS. IN. WG.	TYPE	RPM	WALL OPENING IN.SQ.	DRIVE	HP	ELECTRICAL SERVICE
F-1	EXHAUST	STORAGE	3750	.375	PROP.	860	38 x 38	BELT	3/4	220-3-60
F-2	EXHAUST	EQ. RM.	850	.375	CENT.	850	16 x 16	DIR.	1/4	115-1-60

REMARKS	BASIS OF DESIGN
BACKDRAFT DAMPER	XYZ CO. MODEL NO.100
BACKDRAFT DAMPER	XYZ CO. MODEL NO.100

Figure 9-23. Typical general fan schedule that may be used to describe any fan application.

ROOF EXHAUST FAN SCHEDULE								
SYMBOL	SERVICE	CFM	TOTAL DROP IN.WG	TYPE	FAN RPM	HP	DRIVE	ROOF OPENING
RF-1	TOILET	1200	0.375	CENT.	1460	1/3	BELT	12" x 12"
RF-2	CONF. RM.	500/250	0.250	CENT.	900/450	1/6	BELT	12" x 12"

ELECTRICAL SERVICE	REMARKS	BASIS OF DESIGN
115-1-60	BACK DRAFT DAMPER	XYZ CO. MODEL NO. XX-11-A
115-1-60	"	"

Figure 9-24. Typical specific fan schedule is used to describe Roof Exhaust Fans.

HUMIDIFIER SCHEDULE					
SYMBOL	LOCATION	SERVICE	CFM (S.A.)	CFM (O. A. MIN.)	LVG. %RH
H-1	4th. FL IN DUCT	3rd. FL. W/AC-3	4500	650	40
H-2	3rd. FL IN DUCT	3rd. FL. W/AC-2	5000	550	40

STEAM PRESS. PSIG. MAX.	LB/HR	ORIFICE	DUCT WIDTH	REMARKS	BASIS OF DESIGN
15	27	1/8	46"	- - -	XYZ CO. MODEL AA-25
15	27	1/8	48"	- - -	XYZ CO. MODEL AA-25

Figure 9-25. This type of schedule describes the various types of duct mounted humidifiers used in HVAC systems.

It is considered good practice to include one column entitled REMARKS. This column is reserved for such information that does not really belong elsewhere in the schedule, and sometimes it is used for instructions to the contractor.

The last column "Basis of Design" may or may not be required. This column is used to indicate the quality of the equipment intended or specified by the engineer.

Schedules provide a consistent means of describing equipment and a central location for equipment data. Most importantly, schedules must be legibly lettered to avoid confusion and/or error. On small projects, schedules may be placed on those drawings where the equipment is located. On large projects, there are usually drawings titled "Schedules" and all appropriate schedules are placed on one drawing.

The following is a list of mechanical (HVAC) equipment usually described in schedules:

HVAC EQUIPMENT SCHEDULES

    a. Roof top air conditioner units
    b. Air cooled condensers
    c. Cooling towers
    d. Chillers
    e. Boilers
    f. Fan coil units
    g. Air handling units
    h. Filters

    i. Humidifiers (Fig. 9-25)
    j. Reheat coils
    k. Unit heaters (Fig. 9-22)
    l. Converters
  m. Perimeter radiation
    n. Condensate pumps
    o. Electric dust heating coils
    p. Exhaust fans (Fig. 9-24)
    q. Sound attenuators
    r. Compression tanks
    s. Steam pressure reducing valves
    t. Pumps
    v. Heat exchangers
    w. Hot water generators

Figures 9-22 through 9-25 illustrate some typical schedules.

Chapter 10
Plumbing

In the past, and in some current cases, the plumbing drafter was required to illustrate both the plumbing and fire protection drawings. This has changed, especially in larger firms. Today, there is usually a separate plumbing and fire protection department. This chapter will address plumbing drawings only; fire protection is covered in Chapter 11.

In some firms the plumbing drawings include all plumbing site work. Most firms currently limit plumbing work to that within the building and to a point about 5 to 10 feet outside the building line. There the continuation of all work is picked up by the civil department, which is usually part of the structural department. The extent of plumbing responsibility frequently depends upon office practice. The best division of responsibilities of the various design departments should match the building construction trades.

Plumbing design and drafting are concerned with plumbing fixtures, potable water distribution, and sanitary and storm drains. In medical facilities, the plumbing drawings show all medical gases, distribution piping, terminal devices, and sources which include vacuum pumps, air compressors, gas bottles, etc.

Plumbing is also responsible for compliance with the appropriate plumbing codes. This includes the number and types of fixtures required, based upon occupancy, and the space necessary for the various types and uses of the fixtures. While the architect may have responsibility for space planning, his effort is usually coordinated with plumbing to determine compliance with code requirements.

The level of plumbing drafter and designer involvement depends upon the responsibility assigned to the drafter and designer. Entry-level drafters merely draw from sketches provided by others, designers or engineers, within the department. As the drafter advances within the firm, the degree of involvement increases. In all cases, plumbing drawings must show all work to be performed by the plumbing contractors. Plumbing drawings usually include floor and roof plans for fixture and drain locations, and enlarged drawings for toilet rooms, kitchens, and other such congested areas to show actual piping distribution and connections. On larger and more involved building projects, plumbing drawings also include riser diagrams for potable water and storm and sanitary drains. Plumbing drawings also include typical details of special equipment, schedules, symbols, and abbreviations

Floor plans usually show locations of toilet rooms, floor drains, roof drains, and associated piping. These drawings also show the entrance of potable water, backflow preventers, hot water generators, roof drain piping, and all exit piping to some point beyond the building line.

Since much of the plumbing work consists of supply and return water piping, potable water piping, and storm and sanitary drains, and since this piping may be in the slab, below the slab, and below the floor, various types of lines are used with appropriate letter designations to differentiate among the different uses of these pipes (Fig. 10-1).

Plumbing drawings are, in many respects, an extension of mechanical drawings, particularly the piping portion of the HVAC section. Plumbing systems include water piping, pumps, and valves, and the symbols are similar to those used for HVAC piping. In addition to pumped and pressured piping systems, plumbing drawings include gravity drains for storm and sanitary systems.

Since plumbing drawings are so identified and are used by plumbing contractors, the plumbing drawings include a list of plumbing abbreviations and symbols used on those drawings. The same piping symbol or abbreviation may be used for both HVAC and plumbing systems, therefore each individual pipe line designation, symbol and abbreviation must be explained and described on each drawing. The abbreviation "AD" on HVAC drawings indicates "ACCESS DOOR" while on plumbing drawings "AD" means "AREA DRAIN". Similarly, there may be conflict between pipe line symbols on the two different drawings, but as long as these are explained on both the HVAC and plumbing drawings there will be no confusion between the two.

In addition to pipe line symbols, plumbing fixtures are shown by recognized standard symbols. These include, among others, water closets, lavatories, urinals, shower stalls, electric water coolers, service sinks, etc. (Fig. 10-2).

While many plumbing fixture symbols are readily recognized by their shape (in fact, several drafting templates are available with outlines of the various fixtures), some are merely square, rectangular, or circular in shape. In order to eliminate the possibility of any error, all plumbing symbols are identified by a standard abbreviation. On mechanical drawings the recognized exhaust fan symbol is identified by "EF" adjacent to the symbol, or "EF-1", etc., in the event that there is more than one used on the project. Similarly, plumbing fixtures are shown by standard symbols and letter designations. On large projects, there may be more than one type of a particular fixture; for example, there may be several types of lavatories. In this case, the specific fixture must be identified as "LAV-TYPE A", or "LAV-A". The specific type is then described either by a note on the drawing, in a fixture schedule on the drawing, or in the specifications. In cases where floor plans are drawn at ⅛-inch scale or smaller and the areas are redrawn at an enlarged scale, then the type of fixtures do not need to be identified on both the small-scale floor plan and the enlarged drawing. The identification is usually noted on the enlarged drawing.

1. FLOOR PLANS

Floor plans show locations of all plumbing fixtures, floor and roof drains, various pieces of plumbing equipment, and associated piping to and from these fixtures, equipment, and devices. The piping may start at the curbside service and terminate at a private or public storm and sanitary connection points. On large projects all exterior site work is usually done by the civil engineering design discipline. On small projects, plumbing design may be responsible for all piping

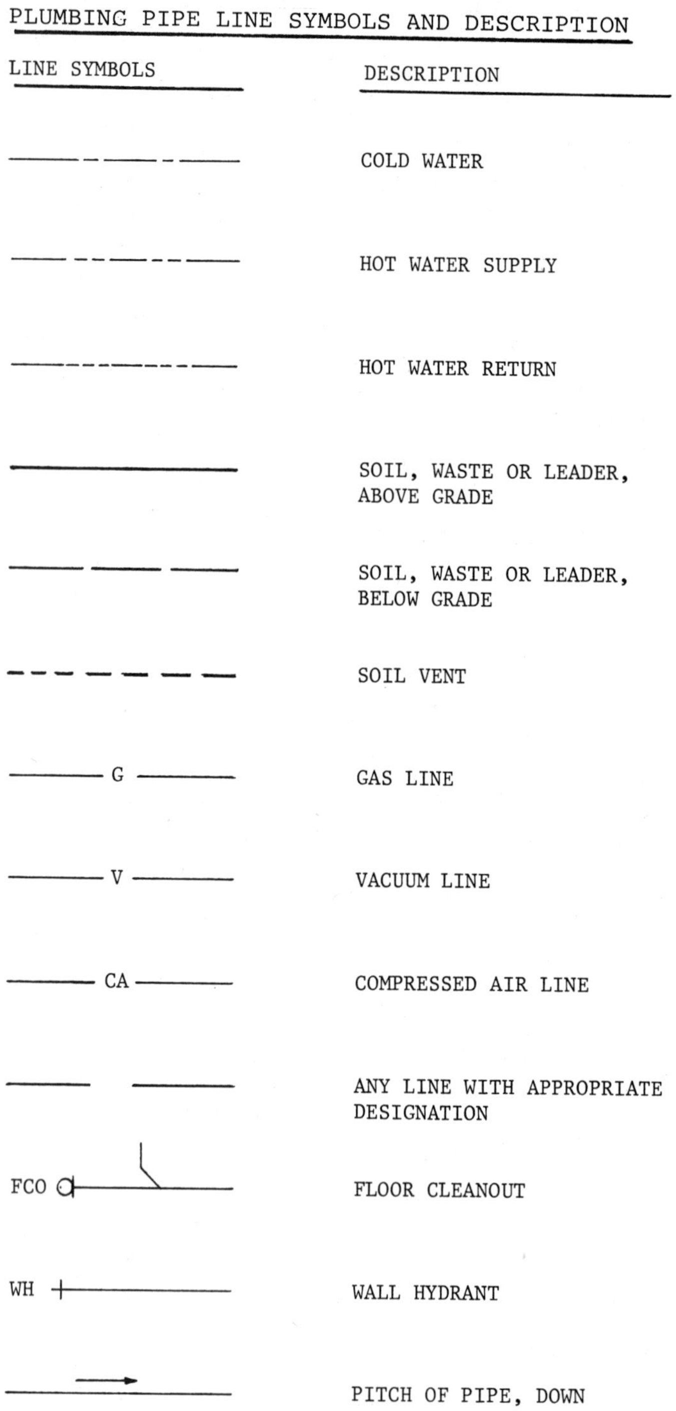

PLUMBING PIPE LINE SYMBOLS AND DESCRIPTION

LINE SYMBOLS	DESCRIPTION
——— — ——— — —	COLD WATER
——— — — ——— — —	HOT WATER SUPPLY
———————————	HOT WATER RETURN
———————————	SOIL, WASTE OR LEADER, ABOVE GRADE
——— ——— —	SOIL, WASTE OR LEADER, BELOW GRADE
— — — — — —	SOIL VENT
——— G ———	GAS LINE
——— V ———	VACUUM LINE
——— CA ———	COMPRESSED AIR LINE
——— ———	ANY LINE WITH APPROPRIATE DESIGNATION
FCO ⊙———	FLOOR CLEANOUT
WH +———	WALL HYDRANT
———→	PITCH OF PIPE, DOWN

Figure 10-1. Common pipe line symbols used on plumbing drawings. Any pipe line can be indicated as long as the appropriate abbreviation is inserted in the line and the abbreviation is described in the symbol list. (From ASA Z32.2.3-1949)

and drainage. When two disciplines are involved it is absolutely essential that the responsibility of each discipline be defined and the point where each terminate and pick up must be established.

The number of fixtures and their general location are dictated by local, state, or national plumbing codes. These codes also establish the size of piping to and from fixtures, the size of drains, and the method of interconnection.

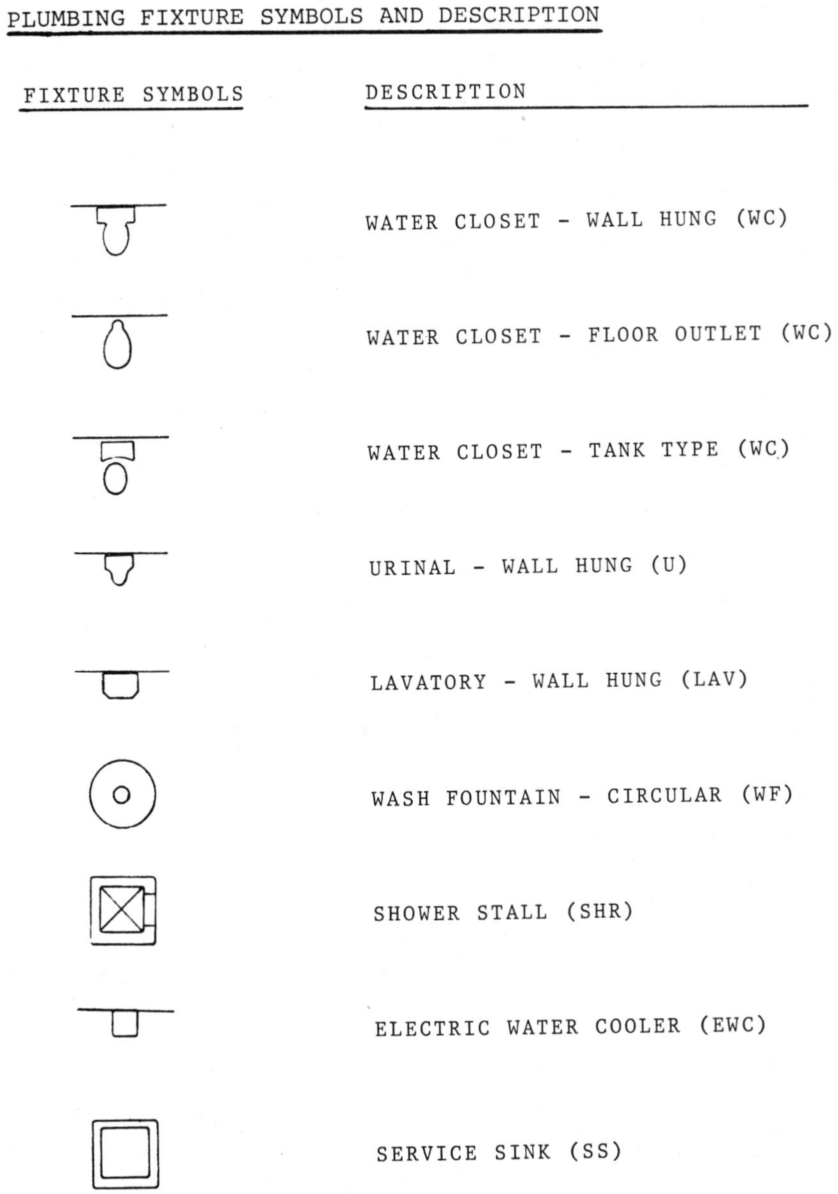

PLUMBING FIXTURE SYMBOLS AND DESCRIPTION

FIXTURE SYMBOLS   DESCRIPTION

WATER CLOSET - WALL HUNG (WC)

WATER CLOSET - FLOOR OUTLET (WC)

WATER CLOSET - TANK TYPE (WC)

URINAL - WALL HUNG (U)

LAVATORY - WALL HUNG (LAV)

WASH FOUNTAIN - CIRCULAR (WF)

SHOWER STALL (SHR)

ELECTRIC WATER COOLER (EWC)

SERVICE SINK (SS)

Figure 10-2. Common plumbing fixture symbols used on plan drawings. Fixture abbreviations are sometimes placed next to the symbols on plan drawings and they are used in place of symbols on isometric drawings. (From ANSI Y32.4-1977)

While architects may select the location of toilet rooms, locker rooms, kitchens, electric water coolers, this work is coordinated with the plumbing department in order to verify code compliance. Once the size and location of these spaces have been determined and the area for the pipe chases sized, the plumbing engineer or designer may develop freehand sketches showing piping distribution for use by drafters for hard-line work on the drawings, or drafters or designers may make preliminary piping distribution drawings for review by the plumbing engineer. In either case, the final drawing shows all fixtures, equipment, and associated piping (Fig. 10-3).

Figure 10-4 is a freehand sketch developed by a plumbing engineer to be hard lined by a plumbing drafter on the plumbing drawing. Since there are several similar types of fixtures used on the project, each fixture is identified by a letter

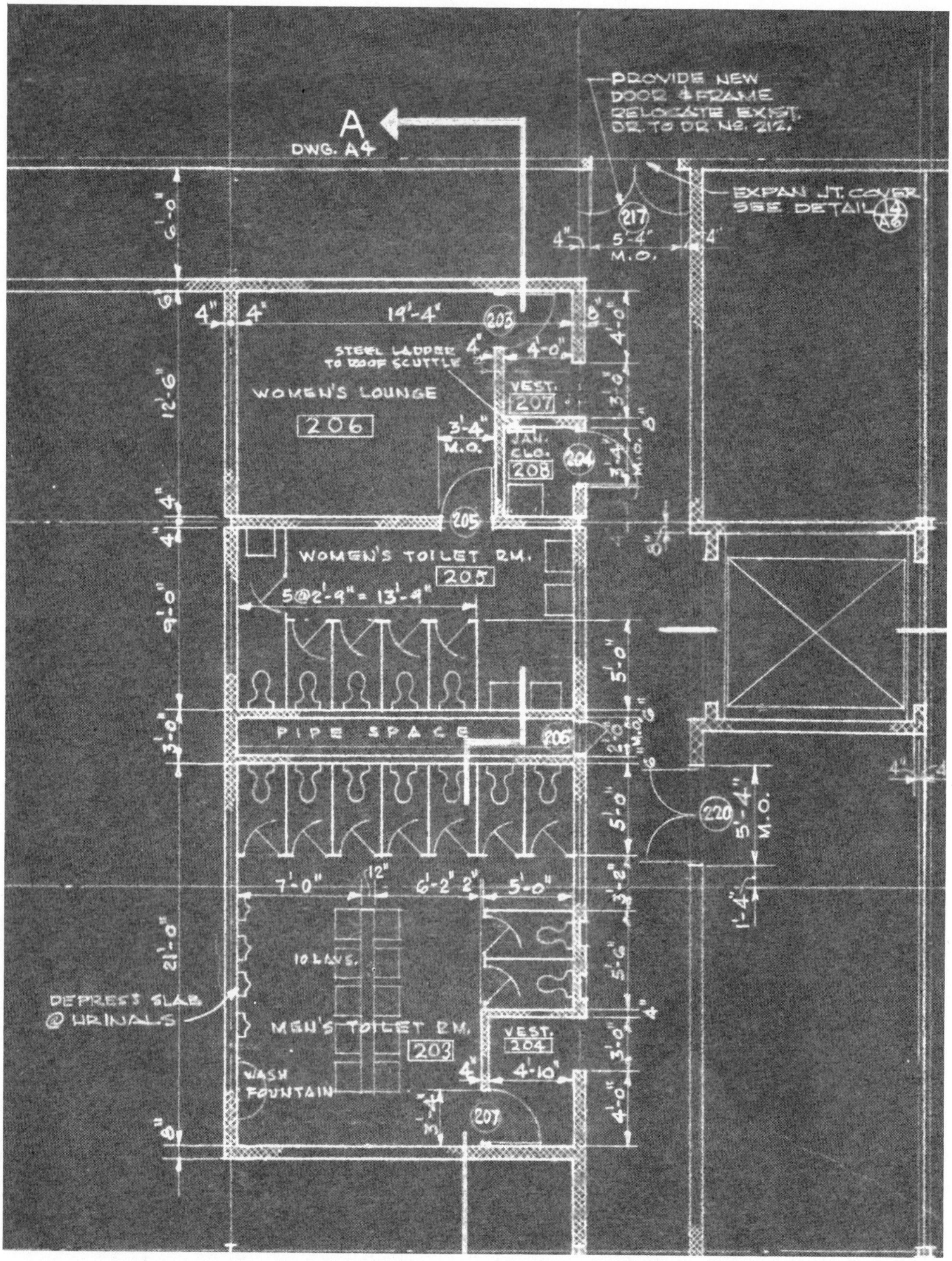

Figure 10-3. Typical architectural floor plan shows the layout and placement of fixtures using typical symbols. Piping is not shown on architectural drawings.

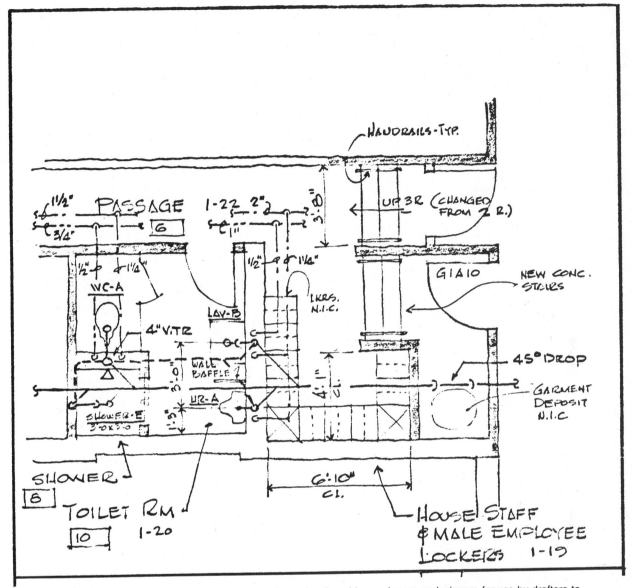

Figure 10-4. Freehand sketches are sometimes developed by engineers or designers for use by drafters to make hard-line drawings.

abbreviation, for example, the water closet is type "WC-A"; the lavatory is type "LAV-B"; the urinal is "UR-A"; and the shower is "SHOWER-E. Also, all cold and hot water lines are shown to the various fixtures, and drain and vent lines are located, and all pipes are sized.

Depending upon the complexity of the project, sometimes separate drawings are made to show the water piping, and the soil, waste and vent piping. Figure 10-5 is a partial water piping plan. The space name is included, fixture types are identified, and pipe sizes are noted. This drawing completely describes the design intent and the contractor should have no excuse for misinterpretation. This drawing shows how well a piping distribution system can be illustrated at a ⅛-inch scale. All appropriate notes are outside the space, with leaders to the point of application. This frees the space from unnecessary clutter, and symbols and pipe runs can be clearly indicated.

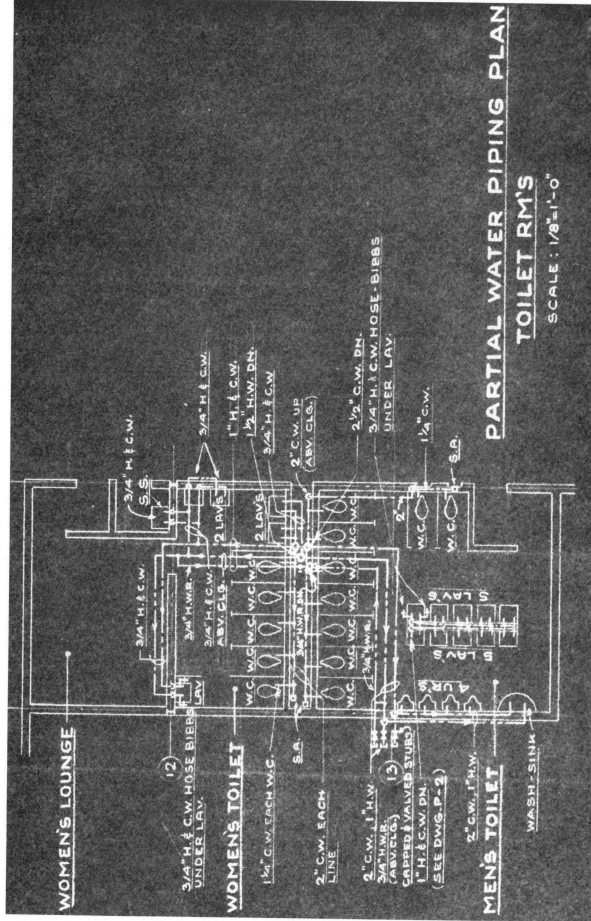

Figure 10-5. Partial water piping plan shows all water piping within the space. All fixtures are identified by abbreviations. Pipes are identified and sized, and the direction of flow is indicated. Typically, soil, waste and vent lines are shown on a separate drawing.

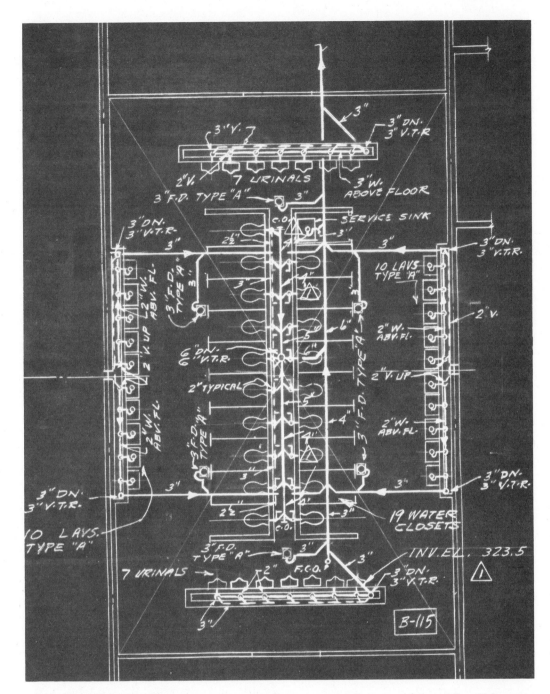

Figure 10-6. Typical floor plan illustrates the method for drawing soil, waste and vent piping. Fixtures are shown graphically by symbols and are also identified, in the case, by groups. The direction of flow is shown and the invert elevation noted.

Figure 10-6 shows soil, waste and vent piping and does not include water piping. That is shown on another drawing. Again, note the details included on the drawing. In this case, each water closet, lavatory, and urinal does not have the abbreviations "WC", "LAV" and "U" at each fixture but does note "19 WATER CLOSETS", "10 LAVS-TYPE A", "7 URINALS." Also floor drains are indicated, and the invert elevation of the soil pipe is noted. Figures 10-5 and 10-6 are partial drawings showing specific areas that could not have been adequately defined on the small-scale drawings for this large building. It is not necessary to make separate drawings for water piping, and soil waste and vent piping on small projects. In the illustrated case, separate drawings were needed because it was impossible to combine both on one drawing and still describe the piping systems properly.

2. RISER DIAGRAMS

In addition to plumbing plan drawings that show all fixtures, equipment, and devices and all interconnecting piping, plumbing drawings also have what is know as *riser diagrams.* These are schematic drawings, usually isometric, but may be two dimensional, that show soil, waste and vent piping, and domestic water supply piping vertically through the building. The actual fixtures are not shown but all piping to and from the fixtures is indicated. While riser diagrams are vertical views of the plumbing systems starting from the building drain and extending through the roof, this view cannot be seen in any elevation cut through the building. These diagrams show the complete system without regard to physical placement of the fixtures, etc., and are not drawn to scale. A single drawing

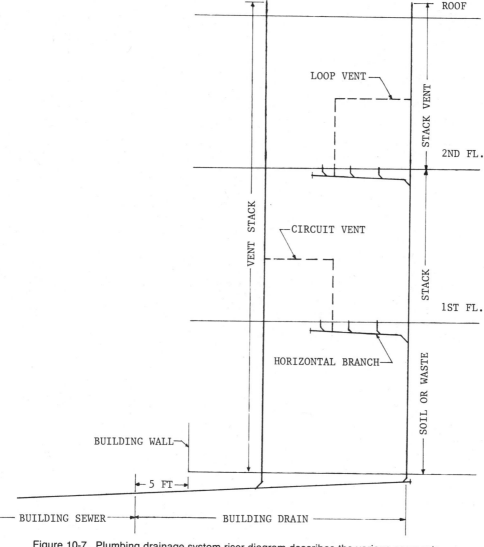

Figure 10-7. Plumbing drainage system riser diagram describes the various segments.

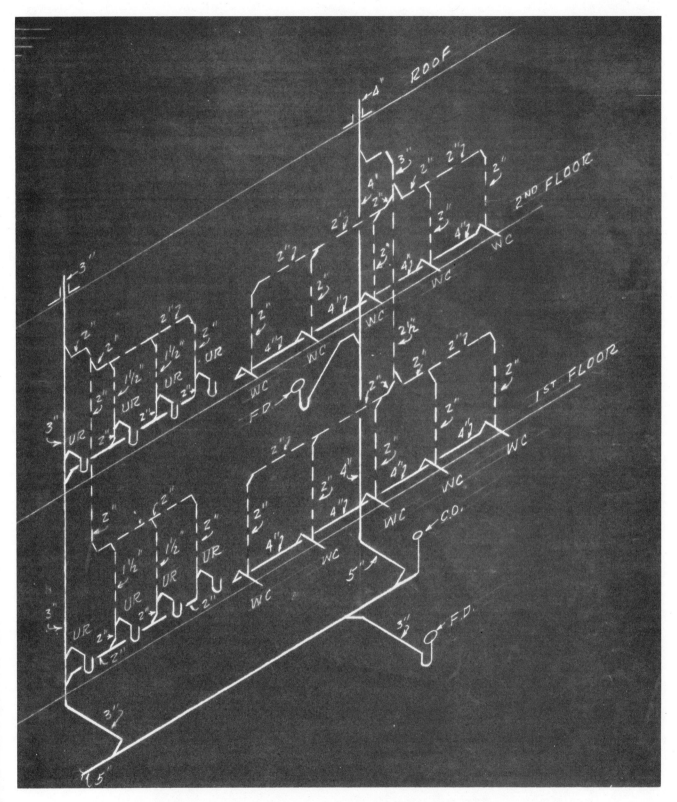

Figure 10-8. Isometric riser diagram shows soil, waste and vent piping for water closets and urinals.

may be made to show soil, waste and vent piping, and domestic water supply piping, or several drawings may be made for different segments of the system.

Some of the common definitions associated with riser diagrams are listed below:

- A building drainage system consists of a building sewer, a building drain, a soil waste stack, horizontal branches (fixture drains), and vents (stack vent and vent stack).
- A stack vent is the extension of a soil waste stack above the highest connected horizontal branch.
- A vent stack is the main vertical stack of a vent system.
- A soil stack is the main vertical stack which receives and conveys the discharge from plumbing fixtures.
- A branch is any part of the piping system other than a main, riser, or stack.

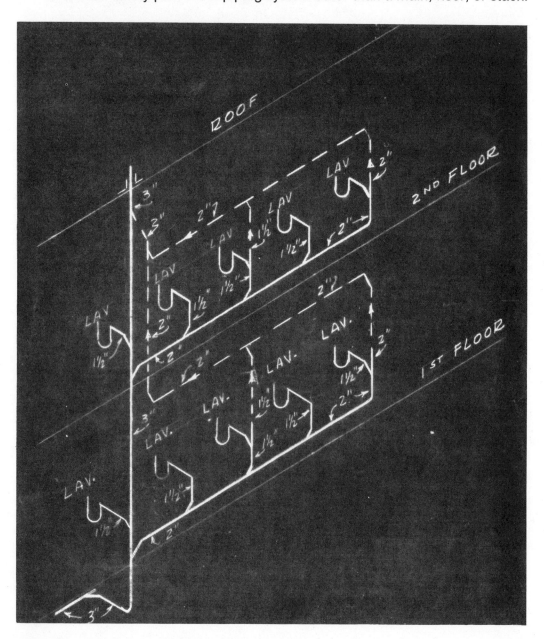

Figure 10-9. Isometric riser diagram for waste and vent piping for lavatories. On small projects, the drainage piping shown on Figures 10-8 and 10-9 may be combined in one drawing.

- A building drain is the lowest horizontal piping of a drainage system within the building which receives the discharge from soil and waste pipes.
- A building sewer is the lowest part of a horizontal drainage system, beginning 5 feet outside the inner face of the building wall, which receives the discharge from the building drain and conveys it to a public sewer or other place of disposal.

Plumbing drawings usually terminate the system at the end of the building drain which, by definition, extends to a point 5 feet beyond the building wall. In many firms, the site work which picks up at that point will show the building sewer in its entirety from the building drain to a public sewer or other place of disposal. Building trades usually follow this division between building and site work on large projects, with the plumbing contractor doing all building work and site contractors doing all site work. On small projects and residences, the plumbing contractor performs all plumbing work. This is the usual practice and is not intended to assign contractor responsibility. Figure 10-7 describes the various terms associated with the soil and waste system.

Figure 10-8 is an isometric drawing of soil, waste and vent piping for water closets and urinals. All fixtures are identified by appropriate abbreviations for

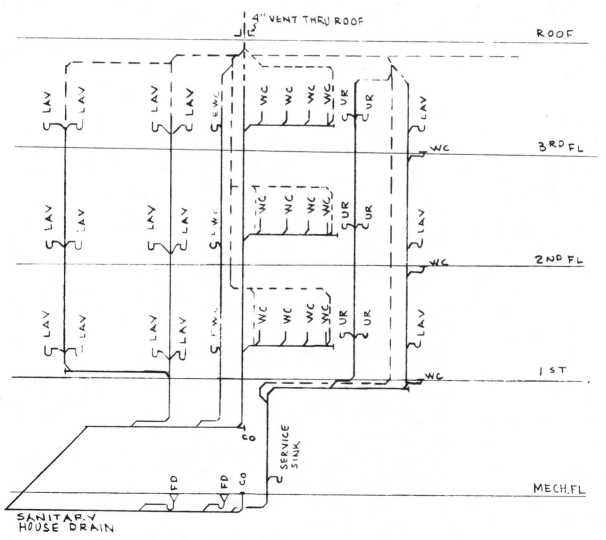

Figure 10-10. Another method for showing a plumbing sanitary drainage riser diagram.

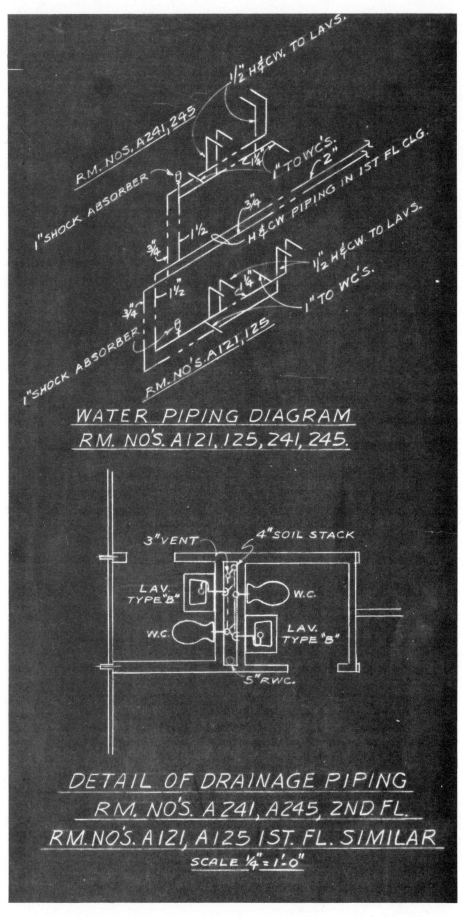

WATER PIPING DIAGRAM
RM. NO'S. A121, 125, 241, 245.

DETAIL OF DRAINAGE PIPING
RM. NO'S. A 241, A 245, 2ND. FL.
RM. NO'S. A121, A125 1ST. FL. SIMILAR
SCALE 1/4" = 1'-0"

Figure 10-11. A ¼-inch scale floor plan drawing shows back-to-back toilet rooms for two floors, one above the other, with its riser diagrams for water piping and sanitary drainage.

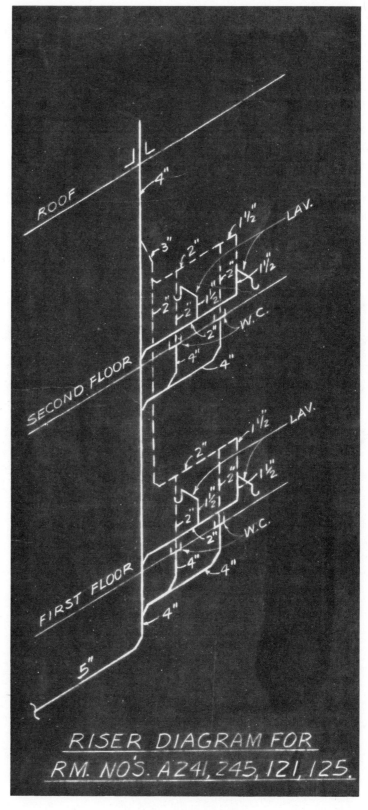

Figure 10.11. *(Continued)*

this two-story building. Floor drains and cleanout are shown, and all piping is completely sized. Solid lines are used for soil and waste piping, and dashed lines are used for vent piping. This riser diagram, taken along with the floor plan for the area, clearly describes the design intent for the two toilet rooms involved. Riser diagrams are not stand-alone drawings, they show connection methods and the floor plans show fixture locations

Figure 10-9 is an isometric drawing for the lavatories in the same two toilet rooms indicated in Figure 10-8. Here again, all appropriate piping symbols are used and all piping is completely sized.

Figure 10-10 is another method of showing the sanitary piping. This type of drawing is easier to make than the isometric and is just as effective. This drawing shows all the plumbing fixtures, stacks, and vents involved, and used the solid line/dashed line convention to indicate the piping, without need to note usage. Because of the type of plumbing contract involved, the piping was not sized. Generally, all riser diagrams fully dimension all of the piping.

Some projects also require domestic water supply and distribution diagrams. These are constructed similar to soil, waste and vent riser diagrams.

Figure 10-11 is a composite drawing showing a floor plan of a back-to-back

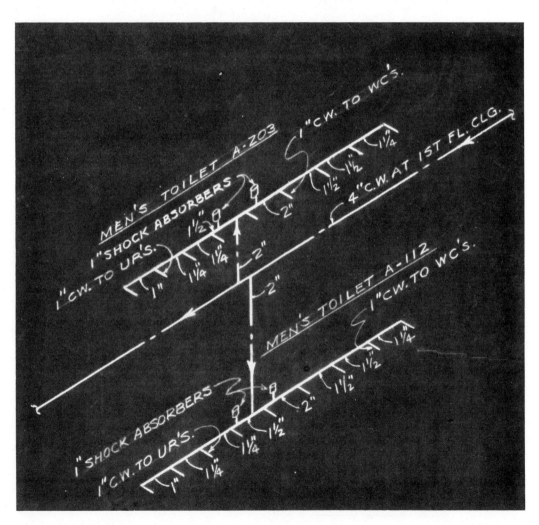

Figure 10-12. Water piping diagram associated with toilet rooms with soil, waste and vent piping shown on Figures 10-8 and 10-9.

toilet room at a ¼-inch scale, a water piping riser diagram, and the associated soil, waste and vent riser.

Figure 10-12 is a water piping isometric drawing for the toilet rooms shown on Figure 10-8 and 10-9, soil, waste and vent piping.

3. ENLARGED DETAIL DRAWINGS

Enlarged detail drawings are as necessary in plumbing design as in any other discipline. In most instances, particularly when floor plans are drawn at ¹⁄₁₆-inch scale, and in some cases when the floor plans are drawn at ⅛-inch scale, enlarged detail drawings are made of the toilet rooms. These are sometimes made at ⅛-inch scale (see Figure 10-5). In these cases, separate drawings are made for the soil, waste and vent piping, and water piping. When these rooms are drawn at ¼-inch scale, both soil, waste and vent piping, and water piping are shown on the same drawing. Office practice and/or space availability on the drawings dictates which method is chosen. Common practice is to use two separate drawings at ⅛-inch scale for large-size toilet rooms, and to use one drawing at ¼-inch scale which shows both drainage and water piping for smaller toilet rooms. In some cases, the choice may be a matter of personal preference.

Plumbing equipment connections are frequently detailed on enlarged detail drawings to indicate the design intent more clearly. These drawings may include backflow preventer connections, sump pump and condensate pump installa-

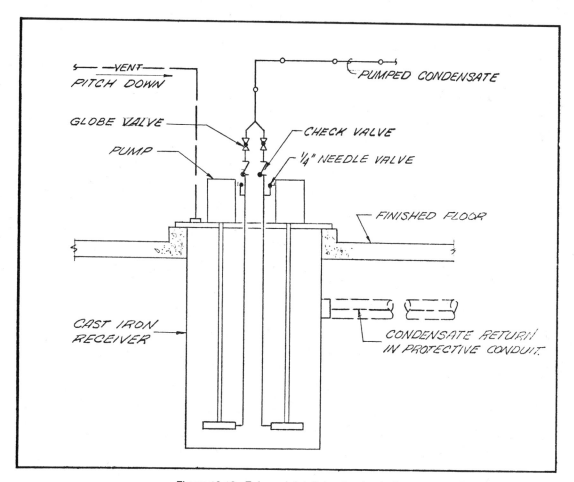

Figure 10-13. Enlarged detail drawing for duplex sump condensate pump.

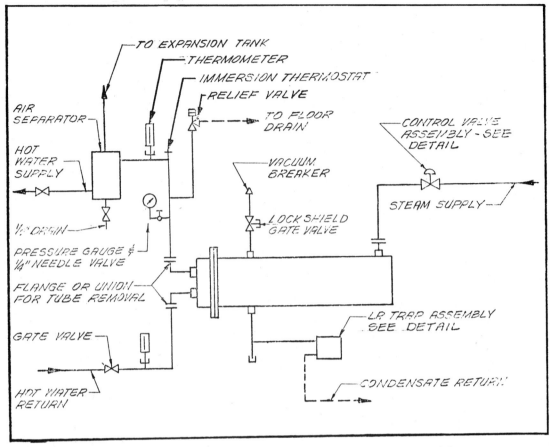

Figure 10-14. Enlarged typical detail drawing shows hot water convector piping.

tions, steam pressure regulating stations, steam pressure reducing stations, hot water generators, hot water convertors, kitchen equipment, medical gas installations and distribution, and other special equipment not normally encountered in routine plumbing work (Fig. 10-13, Fig. 10-14).

Enlarged detail drawings must be properly identified and cross referenced to the appropriate location on floor plans. In many instances, this information is provided by typical details and frequently these typical details are similar to those used on HVAC drawings.

Here also, as in mechanical (HVAC) design, the idea is to provide one typical detail to illustrate connection and installation methods instead of drawing the details in numerous locations on floor plans. There will also be cases where the complexity of piping arrangement cannot be properly shown on small-scale floor plans, and an enlarged (not typical) drawing will be required.

The essential consideration in design and drafting is to not waste time unnecessarily. When one typical or enlarged drawing serves the purpose of describing the work at several locations, use the detail; but remember to note it on the floor plan, i. e., "See Typical Detail on Drawing P-5." Enlarged detail drawings may be typical or specific, depending upon the situation at hand. For example, where several small toilet rooms are scattered throughout the building, one enlarged detail should be used and minimum drafting time should be devoted to the actual location on the floor plan. This is an actual enlarged detaildrawing and not a typical one.

4. SCHEDULES

Schedules are developed for plumbing designs similar to that for other disciplines. These schedules may include plumbing fixtures, drains, fixture connections, and a wide variety of plumbing equipment. The idea is to use a schedule when and where practical to save drafting time, free the drawings from unnecessary notes, and combine in one location all appropriate information about similar types of equipment. On smaller commercial projects and residences, this information is frequently provided in the specifications. The basic format is similar in that, for example, fixtures are identified by an appropriate symbol on the floor plan and that symbol is then described in the specifications.

All plumbing fixtures and equipment must be adequately identified on floor plans by symbols and abbreviations when more than one type is used in the building. These items may be described on the drawings or in the specifications. The method selected should be based on the amount of time (cost) required to describe the item. Drafters generally are not required to select the type of fixture or equipment used; this is done by the designers or engineers. In smaller firms the designers or engineers are also responsible for writing specifications. Larger firms usually have a separate specifications department which has the responsibility for selection and specification of the items.

Schedules, properly utilized, can reduce drafting time and be of valuable service to the building contractor who can quickly and readily identify the various types of similar equipment required for the project. The schedule may show, for example, the different types and sizes of electric water heaters used for the various toilet rooms throughout the building. The contractor will find the appropriate information about *all* water heaters in one location without having to check the entire set of floor plans and this could eliminate the possibility of error. See Figures 10-15 and 10-16.

PLUMBING DRAIN SCHEDULE			
SYMBOL	DESCRIPTION	BASIS OF DESIGN	REMARKS
FD-A	FLOOR DRAIN	XYZ, CO. #2021-A	TOILET ROOM - 120
FD-B	FLOOR DRAIN	" 2020-A	MECHANICAL ROOM
FD-C	FLOOR DRAIN	" 2030-C	KITCHEN
RD-A	ROOF DRAIN	ABC, Co. #4200	ROOF
AD-A	AREA DRAIN	" #4223	REAR DRIVE

Figure 10-15. Plumbing drain schedule.

FIXTURE CONNECTION SCHEDULE						
DESCRIPTION	H. W.	C. W.	SOIL OR WASTE	MINIMUM VENT	SANITARY FIXTURE UNIT VALUE	REMARKS
WATER CLOSET A	- - -	1 1/4"	4"	2"	8	MEN'S TOILET ROOM
WATER CLOSET B	- - -	1 1/4"	4"	2"	8	WOMEN'S TOILET ROOM
URINAL A	- - -	1"	2"	1 1/2"	2	MEN'S TOILET ROOM
LAVATORY A	1/2"	1/2"	1 1/2"	1 1/2"	1	MEN'S TOILET ROOM
MOP RECEPTOR	1/2"	1/2"	3"	3"	3	JANITOR'S CLOSET
SHOWER A	1/2"	1/2"	2"	2"	2	MEN'S LOCKER ROOM
SHOWER B	1/2"	1/2"	2"	2"	2	WOMEN'S LOCKER ROOM

Figure 10-16. Plumbing fixture connection schedule.

Chapter 11
Fire Protection

Fire protection (FP) drafting is similar to that of the other mechanical disciplines; HVAC and plumbing. A fire protection system consists of pumps, valves, piping distribution, sprinkler heads plus gauges, switches, and alarms. Similarly, symbols are used to show the various components of the system as described for mechanical systems.

FP drawings usually include floor plans, enlarged detail drawings, typical details, riser diagrams, a list of symbols and abbreviations, sometimes schedules, and in some cases site drawings. These site drawings are most frequently drawn by civil engineering and then would include all site work such as plumbing and HVAC. Fire protection drafters are concerned with the work within the building and to some point about 5 feet beyond the outside walls. The main concern of the drafter is the floor plan that shows the piping distribution and the location of sprinkler heads. In some cases, all the piping is sized and the location of the heads are fully dimensioned. When a reflected ceiling plan is included in the set of drawings, the location of the heads are indicated on that drawing. Sometimes the area to be sprinklered is merely crosshatched with a note "This area to be sprinklered." In this case, the fire protection contractor makes the drawing as part of his shop drawing submittal which shows the piping distribution and head locations. Pipe sizing and head spacing is dependent upon the degree of hazard as defined by the National Fire Protection Association (NFPA).

There are several types of sprinkler systems that may be required:

- *Wet pipe system*—a system using automatic sprinkler heads in a piping system containing water and connected to a water supply so that water discharges immediately from the sprinkler heads when opened by a fire.
- *Dry pipe system*—a system charged with air under pressure. When a sprinkler head is opened by a fire the water is released from a valve in the system and flows into the piping system and out the opened sprinkler.
- *Preaction system*—a system using automatic sprinklers attached to a piping system containing air that may or may not be under pressure. A supplemental heat responsive device, generally more sensitive than the automatic sprinkler heads, is installed in the same area as the sprinklers. Actuation of the heat responsive device, from a fire, opens a valve that permits the water to flow into the sprinkler distribution piping and to be discharged from any sprinkler that may be open.
- *Deluge system*—a system using open sprinkler heads attached to a piping distribution system connected to a water supply through a valve that is opened by the actuation of a heat-responsive device installed in the same

area as the sprinklers. When the valve opens, water flows into the piping distribution system and discharges from all the sprinkler heads in that area.

- *Halogenated fire extinguishing agent system*—a localized system containing Halon 1301 (bromotrifluoromethane $CBrF_3$). It is a colorless, odorless, electrically nonconductive gas that is an effective medium for extinguishing fires. This system is generally used for areas such as kitchens and computer rooms as well as special manufacturing areas where water systems either cannot be used or are not effective.

There are other systems that may use carbon dioxide, dry chemicals, etc.

There are three classifications of occupancies plus several subgroups that cover the various occupancies:

- *Light hazard occupancies*—includes building occupancies such as office buildings, public buildings, schools, apartments, hotels, hospitals, etc.
- *Ordinary hazard occupancies*—includes building occupancies such as ordinary manufacturing facilities, breweries, canneries, foundries, machine shops, steel mills, warehouses, etc.
- *Extra hazard occupancies*—includes those buildings or portions of buildings housing occupancies in which the hazard is severe as determined by the authority having jurisdiction, such as oil refineries, paint shops, chemical works, etc.

Fire protection piping distribution system is defined as follows: (Fig. 11-1).

- *Risers* are the vertical pipes supplying a sprinkler system.
- *Bulk mains* or *feed mains* are the horizontal pipes supplying risers or cross mains.
- *Cross mains* are those pipes that supply the lines in which the sprinkler heads are installed.
- *Branch lines* are those lines that contain the sprinkler heads and that are connected to cross mains or similar pipes.

Sprinkler pipe sizes are established by code and depend upon the degree of hazard involved (hydraulically designed systems excluded). The following is a partial list of steel pipe sizes required by code and is included for illustrative purposes only. (ALWAYS USE THE LATEST EFFECTIVE ISSUE OF NFPA)

Sprinkler Heads	Light Hazard	Ordinary Hazard	Extra Hazard
2	1 in. pipe	1 in. pipe	1 ¼ in. pipe
3	1¼	1¼	—
5	1½	1½	1½
8	—	—	2
10	2	2	2
40	—	3	3½
60	3	—	—

Also, branch lines should not exceed the number of heads on either side of a cross main as indicated below:

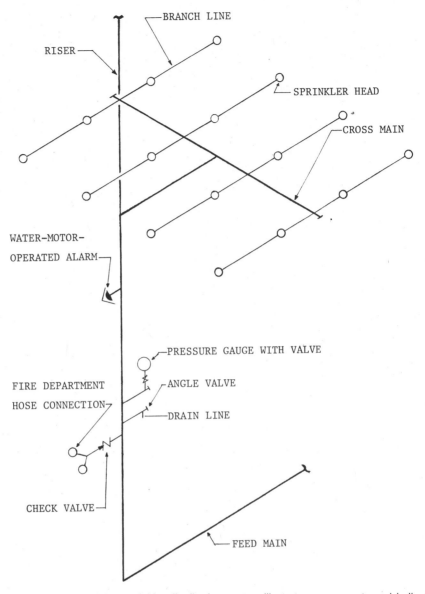

Figure 11-1. Basic fire protection sprinkler distribution system illustrates components and indicates commonly used devices.

Light hazard	8 heads
Ordinary hazard	8 heads
Extra hazard	6 heads

The above pipe sizing and sprinkler head relationship is included for information only, to illustrate general requirements to drafters. Exact requirements may vary depending upon circumstances and the authority having jurisdiction.

1. FLOOR PLANS

Floor plans may or may not be required, depending upon the size of the area involved and the general practice of the office. When major portions of the building require sprinklers, the normal practice is to show the entire floor with head

FIRE PROTECTION PIPE LINE SYMBOLS AND DESCRIPTION

LINE SYMBOLS	DESCRIPTION
—————— F ——————	FIRE LINE
—————— S ——————	SPRINKLER MAIN LINE
— — —S — — —	SPRINKLER DRAIN LINE
————O————	SPRINKLER HEAD
——— DSP———	DRY STAND PIPE
——— WSP ———	WET STAND PIPE
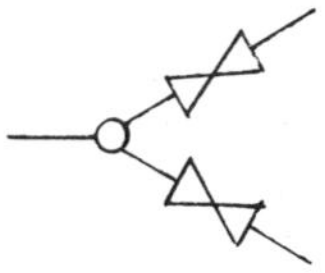	FIRE HYDRANT WITH TWO GATED OUTLETS

Figure 11-2. Common pipe line symbols used on fire protection drawings. (Partially from ASA Z32.2.3-1949)

locations and piping distribution. When only small portions of the floor area are involved, the areas may be drawn at an enlarged scale showing the sprinkler system for that area. These areas may include storerooms, computer rooms, mechanical spaces, and other areas that are considered by code to require fire protection.

Pipe line symbols used on fire protection drawings are similar to the symbols used on other mechanical drawings. The essential difference is that the appropriate abbreviation be used with the pipe, and be noted and defined on the FP symbol list. While there are some standard methods for identifying FP pipe lines, common sense dictates that the letter designation be used. For example, S is used to indicate Sprinkler line, F is used for Fire line. (Fig. 11-2). Floor plans drawn by FP drafters show the entire fire protection system and the basic building outlines identify the area. On large buildings, column lines identify the area; on enlarged scale drawings, room or space identification is used. The one essential consideration is that the intent of the design be fully documented. In areas that do not have a false ceiling, the sprinkler head locations are dimensioned from walls (or partitions) and between the heads.

2. REFLECTED CEILING PLAN

Figure 11-3 is a reflected ceiling drawn by the architect to show all equipment and devices that can be seen from below. The ceiling grid system is used to locate lights, speakers, sprinkler heads, etc., and therefore no dimensions are shown.

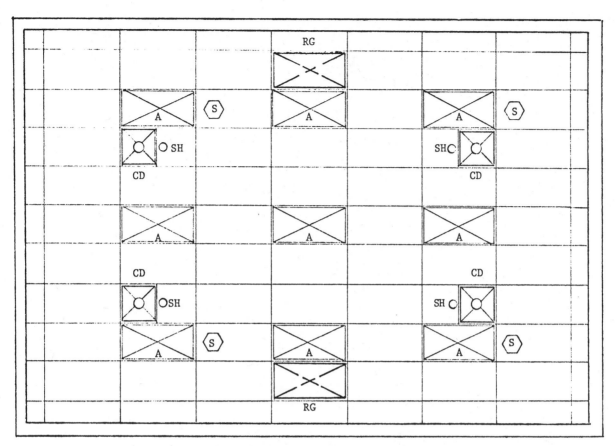

Figure 11-3. Typical reflected ceiling plan drawn by architects shows, along with other devices, the fire protection sprinkler heads, identified by a circle and the letters "SH".

Figure 11-4 is a reflected ceiling plan that is part of the fire protection set of drawings. The sprinkler heads are identified by a circle within a circle to differentiate them from the lighting fixtures used in the space along the right hand side of the drawing. Other lighting fixtures are noted as rectangular shapes. In this case, the location and spacing of the sprinkler heads are dimensioned. This ceiling consists of 24-inch square ceiling tiles glued to a suspended sheetrock ceiling. The fire protection department felt that for the required placement of all sprinkler heads, complete dimensioning was necessary. This is a special case and more often than not reflected ceiling plans are not made by the FP department and the reflected ceiling plan does not include dimensions. This example was included to show what may be required.

In cases where uniform coverage in intended and the sprinklers are so spaced, sprinkler head locations are not dimensioned. Areas where automatic sprinklers are not required are so noted to indicate design intent clearly.

Figure 11-5 is a partial sprinkler floor plan. It shows the entire distribution system from the underground supply to the sprinkler heads. The design calls for a 6 inch supply, a 6 inch riser with water flow device and drain, a 6 inch feed main and various sized cross mains reduced in the direction of lesser flow requirement. The branch lines are not sized since the degree of hazard is defined in the specifications and the contractor will size the lines according to code requirements.

Two different types of sprinkler heads are required as indicated by the two different sized circle symbols. The actual types are identified on the symbol list.

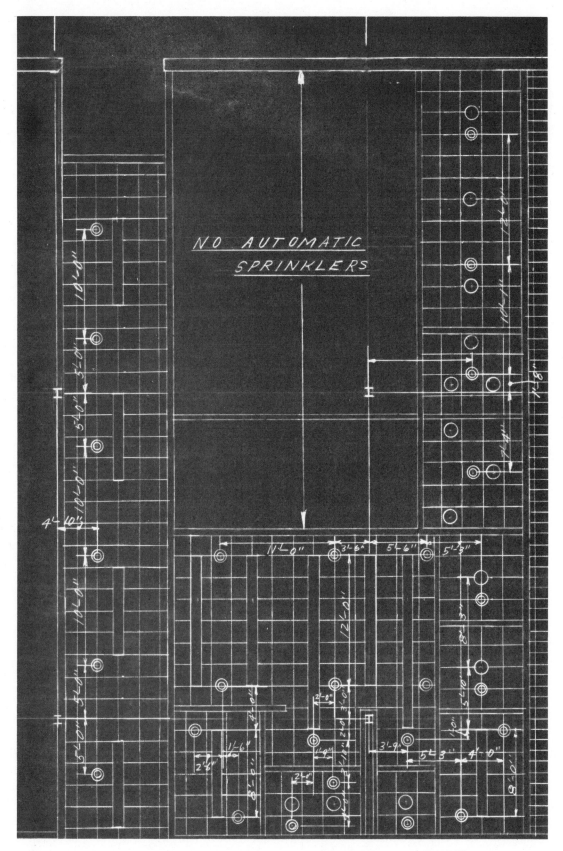

Figure 11-4. A typical reflected ceiling plan drawn by the FP discipline and included in the FP set of drawings.

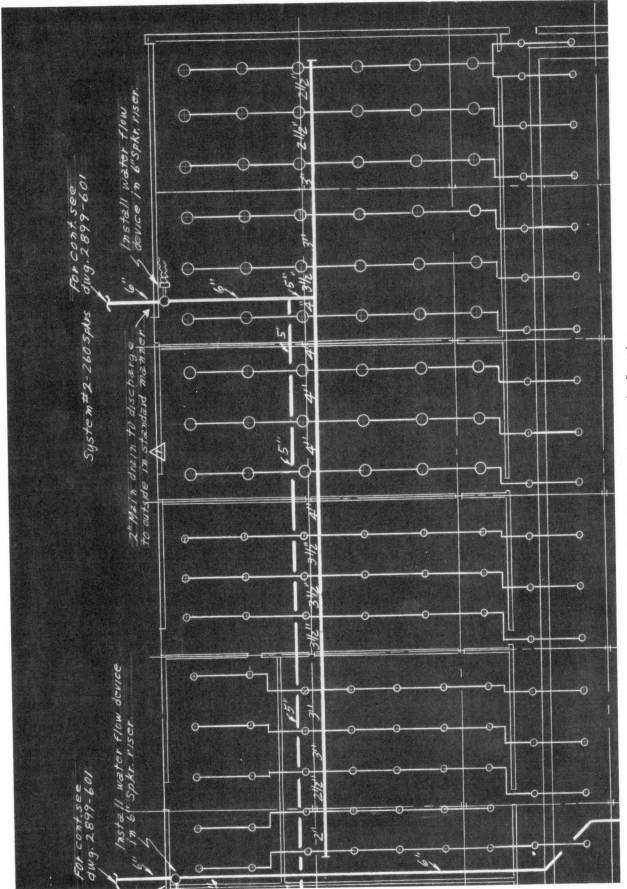

Figure 11-5. Partial fire protection floor plan.

197

Sprinkler head locations and spacing are not indicated on drawings for uniform distribution and coverage.

Figure 11-5 is a typical type drawing developed by the fire protection department. It is part of a fire protection set and shows only fire protection related equipment and devices.

Fire protection drafting is similar to mechanical drafting, i. e., HVAC and particularly plumbing. Mechanical drafting illustrates piping systems from some water source to the terminal devices; in FP, it is from the service to the sprinkler heads; in plumbing, it is from the service to the fixtures; and in HVAC, it is from pumps around and through equipment and back to the pumps. All of these discipline drafters use a wide variety of symbols to illustrate the piping, pumps, valves, fixtures, devices, etc. All three depend upon their own symbol list to describe the symbols used by the specific discipline.

In years past, and to some extent even today, fire protection drawings were made under the direction of the plumbing department. More recently, as fire protection systems have become more sophisticated and complicated, the trend is toward an independent fire protection department.

Fire protection is more than just sprinkler piping, it is outside screw and yoke (OS&Y) and post indicator valves with tamper switches, flow switches and alarms, pressure gauges, gate, globe and check valves, fire pumps and jockey pumps, wet and dry stand pipes, fire hose cabinets, fire extinguishers, all designed and installed in accordance with NFPA standards and any other authority having jurisdiction for the work.

The fire protection drawings reflect the intent of the specifications and are coordinated with the electrical department. While sprinkler system design and drafting is by the fire protection department, the alarm and control system is usually the responsibility of the electrical discipline.

Chapter 12
Structural

Structural drafting is somewhat similar to architectural drafting, much more so than to other engineering discipline drafting. Also, structural drawings are different from both architectural and engineering drawings, i. e., structural drawings do not have floor plans and roof plans, as such. Structural drawings have foundation plans, floor framing plans, and roof framing plans. Framing plans show the structural support for the floors and roof, and show the structural construction of the floors and roof. Structural drawings depend to a great extent on details and schedules. The details show items such as typical framing bays, column and beam connections, framed openings, reinforcement used in concrete structures, pedestals, footings, foundations. Schedules are used quite extensively on structural drawings and may include schedules for columns, pedestals, concrete slabs, footings, retaining walls.

There are three basic types of structures used in building construction, i. e., wood, steel, and concrete. These systems may be used independently or in some combination.

Wood structures are used primarily for residential type of buildings and frequently the architect will provide the framing plans and details. When wood is used in commercial and industrial buildings, the framing plans are usually made by the structural designers and drafters.

Steel and concrete structures are used in commercial, industrial, and institutional building (CII), and structural engineers are responsible for the design. Foundation and framing plans are drawn to scale, generally matching the scale used by the architect, and are basically diagrammatical in nature. Beams and joists are represented by a single line, columns by the appropriate symbol, and footings and foundations by squares, circles, or rectangles to indicate the shape and general size. Construction details and dimensions are fully described either in detail drawings or in schedules. Since many items, such as interior columns, interior bays, interior footings, etc., are usually identical, one typical detail is used to describe the design intent, with reference to locations on the foundation and framing plans. This procedure eliminates the need for extensive detailing on the plan drawings.

Structural drafters must be capable of producing quality details that are accurately scaled and correctly dimensioned. Also, structural drawings must be thoroughly coordinated with drawings developed by the architect and other engineering disciplines. Structural drawings must be compatable with the architectural design and must provide support for and access to electrical and mechanical equipment.

Depending upon the size of the project, structural drawings may show soil conditions and test boring locations, or this information may be included on site drawings developed by the civil engineering discipline. In most firms the civil engineering function is part of the structural department.

While structural design is performed by engineers, drafters and designers need to be able to interpret the design and develop the appropriate details. This detail may include reinforcement in concrete structures and connection methods in steel structures. Concrete structures require a variety of reinforcing bar (rebar) sizes that are used in footings, foundation walls, grade beams, columns, and flat slabs. Rebars are identified by numbers related to the diameter of the bars, i. e., the number represents 8th of an inch:

Number	Diameter (Inches)
3	.375
4	.500
5	.625
7	.875
8	1.000

Welded wire fabric is often used in slabs on grade and in some cases in floor and roof slabs. A typical designation: 6 × 6-8/8 describes the spacing and size of the wire. The first two numbers indicate the longitudinal and transverse spacing in inches respectively and the latter two specify the wire gauge. Therefore the above designation means that #8 wires are spaced at 6 inches in each direction.

Foundation walls are usually keyed to the footings, and the key method must be detailed. In some instances, water stops are required and must be properly indicated. Expansion joints, when required, must be indicated and detailed, and construction joints must show the type and location. It is quite obvious that concrete structure detailing is somewhat involved and the structural designer and drafter must learn how to develop the appropriate detailing techniques.

Steel structures use a variety of structural shapes, and these members are connected by one of three methods; welding, riveting, or bolting. Some of these connections are performed in the shop and some in the field; where they are made must be noted on the drawings. Again, while engineers perform the necessary design, drafters and designers develop the drawings. Plans are diagrammatical in nature, and detail drawings provide the essential information describing construction and installation methods.

1. FOUNDATION PLANS

Figure 12-1 is a basic foundation plan for a building as developed by some consulting firms. It shows the location of the footings and foundation walls, the bay spacing and overall building dimensions, and the necessary detail information is included in the attached notes:

CONCRETE NOTES

1. All concrete to be 2500 psi at 28 days.
2. All foundation walls to be 1'-0".
3. Footings to extend a minimum of 6" beyond foundation.
4. ½" × 5" bituminous filler all around between slab and foundation wall.

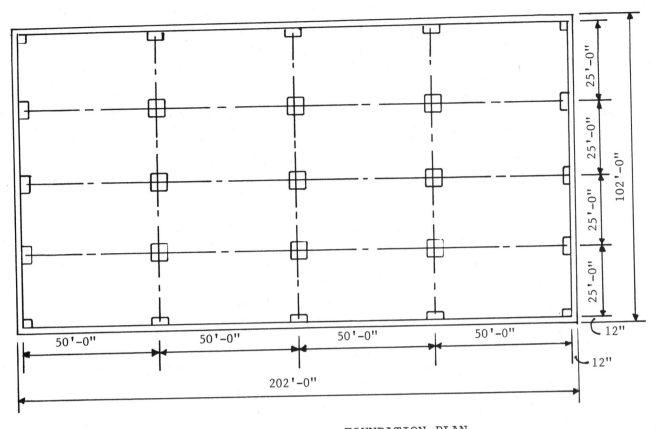

FOUNDATION PLAN

SCALE: 1" = 20'-0"

Figure 12-1. Typical foundation plan includes the location of all footings and is fully dimensioned.

5. Footings to be 6'-0" × 6'-0" × 2'-6" deep, or 6'-0" × 6'-0" × 1'-0" deep with #4 rebars at 6/6.

6. Slabs to the 5" thick with 6 × 6-10/10 welded wire fabric. Smooth steel trowel finish.

7. Top of slab elevation is 0'-0".

8. Top of foundation and pier elevation is -0'-6" minimum.

Figure 12-2 is another type of foundation plan. Since this structure required different type and sized footings, these are identified by letter notation and are described in the footing schedule. Also, two section cuts are indicated, and since they are identical, one detail marked "Section W-28" provides the necessary construction details.

In most cases, foundation plan drawings are rather simple to make. The amount of detail included depends upon office practice. Most foundation plans are dimensioned, but some are not, again depending upon office practice. In either case, the scale used is the same as that used by the architect.

Where different type and size of footings are required, the simplest method is to identify the various footings by letter notation and to use this mark on the footing schedule where the size and reinforcement is fully described. Usually, the quality of the concrete is noted in the specifications or in notes on the drawings.

Foundation plan drawings should be as simple as possible, identifying only the essential elements. Basically, they are grid drawings showing by center lines

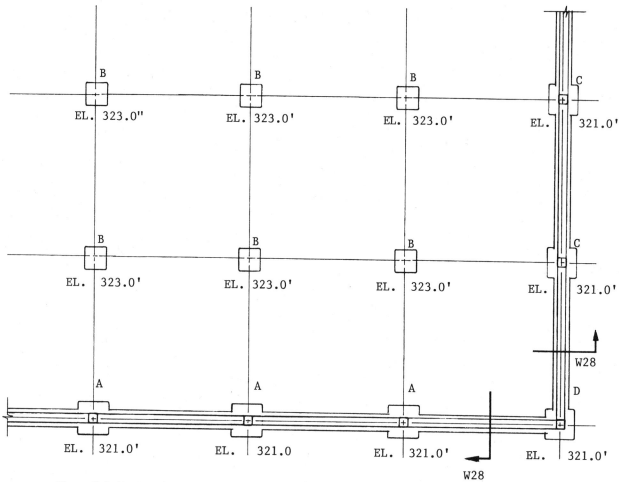

Figure 12-2. Partial foundation plan shows appropriate elevations and locations of footings and locations and view directions of section cuts.

the locations of footings, foundations, piers, walls, etc., and appropriate notations. The grid is to scale and the building outline may or may not need to be dimensioned. The title block is usually used to indicate that the drawing is the foundation plan, floor framing plan, or roof framing plan. In cases where the foundation plan does not fill the drawing, a note under the outline describes it as a foundation plan, etc. The scale at which the plan is drawn is always indicated, either in the title block or under the plan.

2. FRAMING PLANS

Framing plans are identified by where the framing is intended, i. e., floor framing plan or roof framing plan. Framing plans are rather simple and straightforward. Steel framing plans generally show the column locations and interconnecting framing members. In some cases, the column types are noted at each location, and in others the columns are described in column schedules. Column lines are usually noted on the architectural drawings, and this same notation is used on structural drawings. For example, a letter notation is used in one direction and numbers in another. In this manner, any specific column can be located at the intersection of the two column lines. For example, Column C-4 indicates the

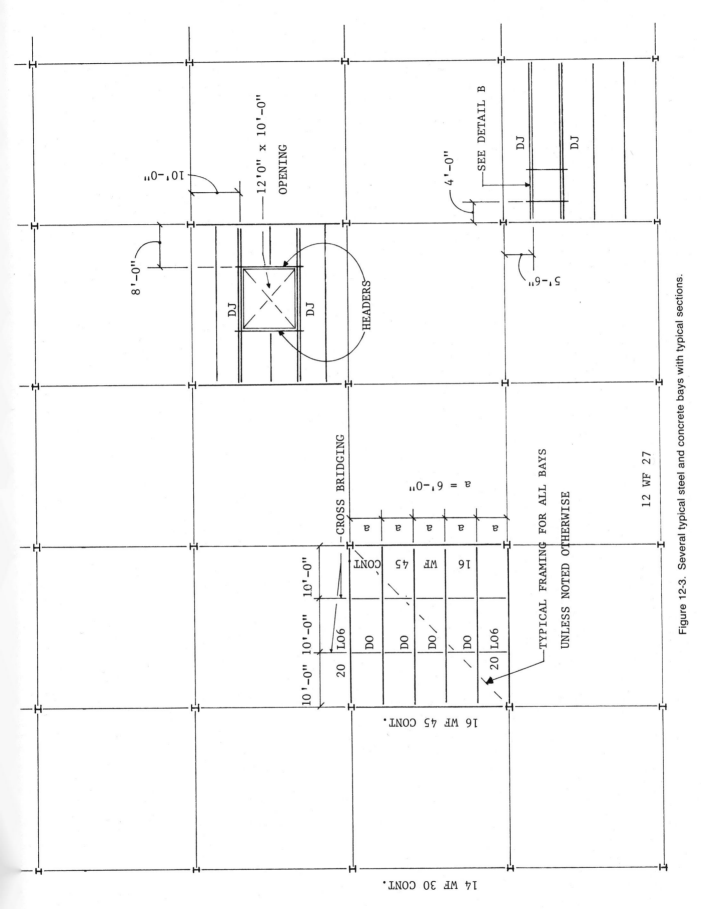

Figure 12-3. Several typical steel and concrete bays with typical sections.

column that is at the intersection of Column Line C and Column Line 4. There can be only one column C-4.

Figure 12-3 is a partial steel roof framing plan. In order to simplify this plan, only three bays show the framing required. A typical bay framing plan is shown and is to be used for all bays except as indicated by the note "Typical Framing For All Bays Unless Noted Otherwise." This method eliminates the need for extensive drafting, particularly where some one hundred or more bays are involved. This one bay detail shows the beams, bar joists, cross-bridging, identifies the sizes of the steel members, and indicates spacing dimensions. No further drafting time is required for any other typical bay.

Two other bays are illustrated on this partial plan. One is for a framed opening for mechanical equipment. It shows the size of the opening and its location with respect to an adjacent column. Double joists (DJ) are required along one side and headers along perpendicular side. The second bay is used to support mechanical equipment suspended below and the note "See Detail B" indicates that additional essential information may be found in that detail. This bay contains less information than the one used for the framed opening bay because the balance of the information is provided in the detail.

This technique is commonly used on framing plans because it saves drafting time and eliminates unnecessary clutter on the drawing that generally serves no useful purpose. Column lines are identified as described above, usually along the top of the drawing and along the right-hand side. Column type and size may be identified on the framing plan or included in the column schedule.

Figure 12-4 shows four different methods for making typical bay framing plans. The upper two are for steel buildings and the lower two are for concrete structures. The typical framing plan would be for the entire floor or roof of the building, again as noted above, "Except As Noted..." All columns are shown and all interconnecting steel or concrete is noted. Steel structures may show all members for the entire plan, since this requires only interconnecting lines. Most importantly, the members must be noted, at least in the typical bay with lines drawn to indicate areas included. Some drafters show the size of one member at the top of the sheet and use the letters "DO" to indicate ditto, meaning that all steel below the one identified is the same. These various methods are drafting shortcuts and should be used to reduce drafting time. Also, when sizing steel on a framing plan, care must be exercised to eliminate possible error by inadvertently writing in incorrect numbers. Using "DO" reduces the possibility of error.

The same basic principle applies to concrete framing plans. Imagine the amount of drafting time that would be required if all bays on a framing plan were drawn as shown in either of the two lower illustrations in Figure 12-4. Common practice is to show a typical bay with appropriate sections. Remember, structural drawings rely heavily on details where a particular section is cut only once and one detail will suffice.

Figure 12-5 shows several common steel structure details. A most important consideration in steel construction is the method of interconnecting the various steel framing member. In concrete buildings, rods and bars are used to interconnect the various members such as footings, foundations, slabs, column beams, etc. In residential wood frame construction, this is generally accomplished by nails; in commercial and industrial wood construction framing, nails, screws, bolts, and rods with steel plates are used. Steel framing members must also be fastened together, and as with other framing systems, the steel must also be anchored to the earth. Piles, piers, and footings are usually part of the

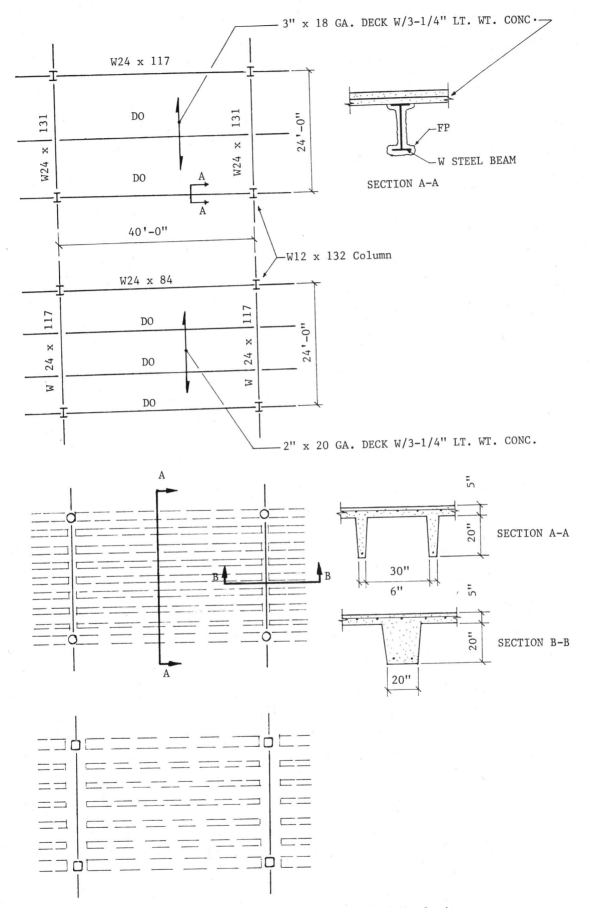

Figure 12-4. Partial roofing plan indicates typical interior bay framing.

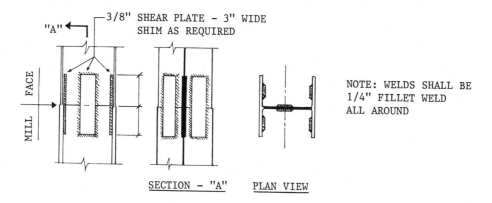

TYPICAL COLUMN SPLICE DETAIL

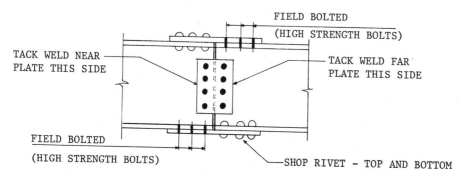

TYPICAL CONTINUOUS MEMBER SPLICE DETAIL

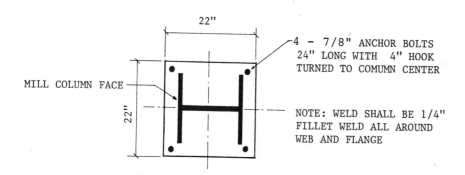

TYPICAL BASE PLATE DETAIL

Figure 12-5. Typical steel framing member details.

substructure of steel buildings, and the connection at this point is frequently accomplished with some type of base plate. This base plate also serves as a platform for the steel superstructure.

A Typical Column Splice Detail is necessary in multi-story buildings where columns are continuous and are spliced. The size and length of the columns are indicated on column schedules, as is the location of the splice. Sufficient details are necessary to inform the contractor exactly how these splices must be made.

Some buildings are designed with continuous horizontal structural members and these members also have to be spliced. In these cases, a Typical Continuous Member Splice Detail is required. When some work is to be performed in the shop and some in the field, this must be so indicated. Some offices completely detail the bolts, while others use the simplified technique illustrated in Figure 12-5. This method is satisfactory in most cases, and it is not necessary to waste drafting time drawing the bolt heads and the nuts. Simplified drafting saves time.

There are several methods for showing Typical Base Plate Details. It is necessary to show the design intent, and the simplest method that satisfies this purpose is adequate. The thickness of the base plate may be noted on this detail or it may be included in the column schedule or in the specifications.

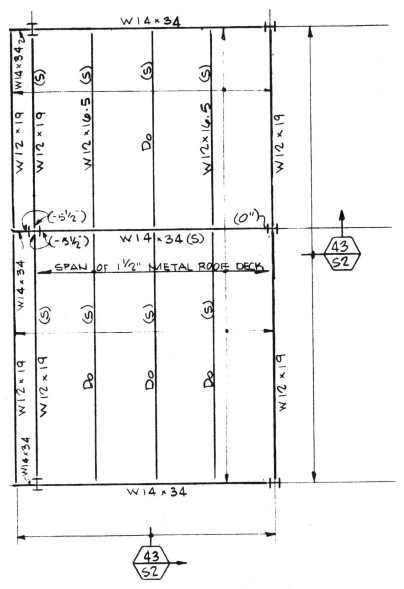

Figure 12-6. Partial steel roof framing plan.

Framing plans must be as complete as necessary to describe the design intent fully. Framing plans, like foundation plans are drawn to scale, and most frequently to the scale established by the architect. All structural members must be adequately illustrated. In most cases, a single line is used to show the horizontal member regardless of its size and shape, and this information is noted as required. Where typical bays can be used to advantage, it is not necessary to show that detail in each bay. Special framing areas must be shown and noted with dimensions where appropriate. Where additional information is required, it may be provided in additional details elsewhere on the drawings (Fig. 12-6).

3. DETAILS

Concrete structures frequently require numerous details to describe construction methods fully that cannot or are not shown on plan drawings. Figure 12-7 shows three examples of typical concrete structure details. In many instances, one typical detail is included with a common note "See Schedule", which indicates, as in Figure 12-7, that some dimensions are not the same for all cases. This information is then provided in the appropriate schedule. It is not necessary to draw several beam, joist, or bridging details when one adequately illustrates the concept and the schedule provides the additional dimensions.

In some cases, plans indicate section cuts where necessary. These locations are indicated on the plan with a note indicating the specific cut. These may be noted as in Figure 12-7 as "Section W28," or as frequently used by some design firms "Section 5 on Drawing 8" using the architectural notation with the number 5 over the number 8 in a circle (see Figure 6-21).

Details are drawn to scale and the scale may also be indicated as for "Section W28" or may be omitted as on the details for Pedestal Type and Thickened Slab (Fig. 12-8).

The Stair Framing Detail fully describes the method of construction. Floor elevations are indicated, slab thickness is dimensioned, and rebar requirements are detailed. The number and size of risers and treads are indicated and the overall dimensions noted (Fig. 12-9).

Appropriate details are an essential part of structural drawings, and drafters and designers need to be able to make accurate drawings at a scale sufficiently large to permit clear detailing. The scales used for detailing are independent from the scales used on plan drawings. The intent of details is to provide the necessary information so that the structure can be constructed. The type and amount of information included on details depend upon the complexity of the construction. The information should be adequate to the needs of the detail. For example, the quality of the concrete was not defined on any of the above details because this information is in the specifications or in notes on the drawing. Never draw or note more than necessary because, first, it is a waste of time and second, the more places the same information is included the greater the opportunity for error. Some information can be more easily included in schedules or specifications, and that is where it belongs. Never clutter details with redundant information, since this distracts from the quality of the detail.

4. SCHEDULES

Structural schedules are like any other discipline schedule, they are a simplified method for combining similar type information in one location. Since plan draw-

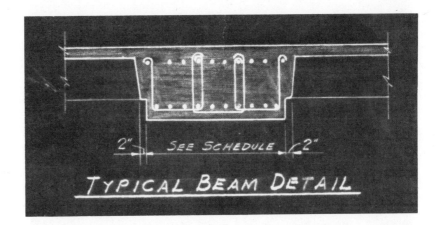

TYPICAL BEAM DETAIL

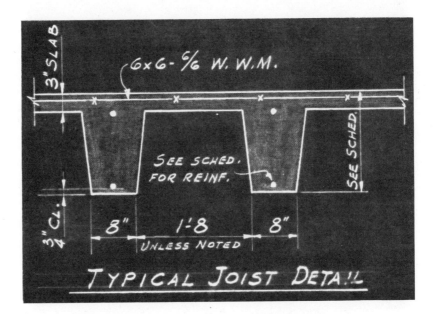

TYPICAL JOIST DETAIL

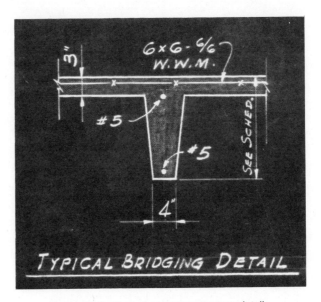

TYPICAL BRIDGING DETAIL

Figure 12-7. Typical concrete structure details.

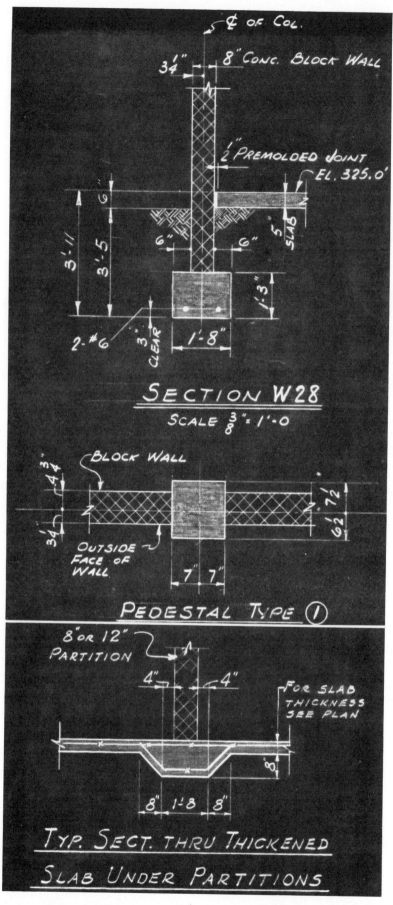

Figure 12-8. Common concrete section cuts and typical details.

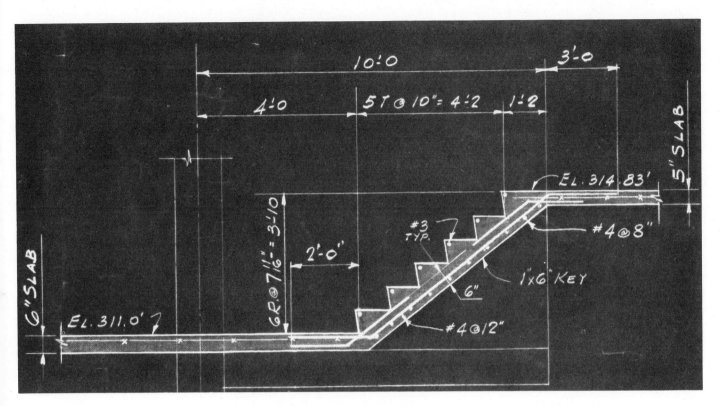

Figure 12-9. Stair framing detail.

ings are generally diagrammatical in nature and do not contain all of the information required for construction, this detailed information must appear somewhere. The choice then is between detail drawings and schedules. Most frequently, detail drawings are typical for a particular type of application and the necessary specifics are described in schedules. Residences and small building projects do not necessarily require schedules as this type of information is frequently more easily developed in notes on the drawings or in the specifications. Large building projects are most often best described through the use of schedules. Some of the more common schedules include information about columns, footings, concrete slabs, pedestals, and retaining walls. There may be more, but this depends upon the scope of the project (Figs. 12-10 through 12-14).

Structural schedules are in many ways similar to the schedules described for architectural, electrical, and mechanical disciplines. Whenever there are several similar items that have similar characteristics, a schedule could be developed to display the necessary information. At this point review all schedules included in earlier chapters to understand the basic mechanics for the development of schedules.

Drafters and designers should realize that plan drawings are pictorial and are used to illustrate the basic building. Details describe specific construction methods for particular areas, and schedules provide a central location for information about similar items and equipment. Specifications are used to describe means and methods of construction and the quality of material. Very frequently, on small projects and for residential construction, the technical section of the spec-

COLUMN SCHEDULE							
COL. SUPPORT'G \ COLUMN NO.		F-3.6	F-5	F-6	G-3	G-5	G-6
ROOF							
4th FLOOR		12 WF53	12 WF50	12 WF53	12 WF40	12 WF40	12 WF40
3rd FLOOR	COL. SPLICE 3'-0"						
2nd FLOOR		12WF 85	12WF 85	12WF 85	12WF 85	12WF 85	12WF 85
1st Floor							
BASE PLATE	SIZE N-S (in)	22	22	22	22	22	22
	SIZE E-W (in)	22	22	22	22	22	22
	THICKNESS	1 1/4	1 1/2	1 1/4	1 1/4	1 1/2	1 1/4
LEVELING PLATE		3/8	3/8	3/8	3/8	3/8	3/8
GROUT		7/8	7/8	7/8	7/8	7/8	7/8
PEDESTALS	TOP ELEVATION	-20"	-8"	-32"	-20"	-32"	-32"
	SIZE N-S (in)	24	24	24	24	24	24
	SIZE E-W	24	24	24	24	24	24
	VERT. REINF.	8-#7	8-#7	8-#7	8-#7	8-#9	8-#7
	TIES @ 12" C.C.	#3DT	#3DT	#3DT	#3DT	#3DT	#3DT
FOOTINGS	SIZE N-S (in)	7'-0"	9'-0"	7'-0"	5'-0"	6'-3"	6'-0"
	SIZE E-W	4'-0"	4'-0"	4'-0"	5'-0"	6'-3"	6'-0"
	DEPTH	1'-0"	2'-0"	2'-0"	1'-0"	1'-2"	1'-1"
	REINF. N-S	8-#6	8-#6	8-#6	6-#5	9-#5	9-#5
	REINF. E-W	8-#6	14-#6	10-#6	6-#5	9-#5	9-#5
TOTAL LOAD ON TOP OF FOOTING		157k	167k	157k	134k	219k	209k

Figure 12-10. Column schedule includes all appropriate information.

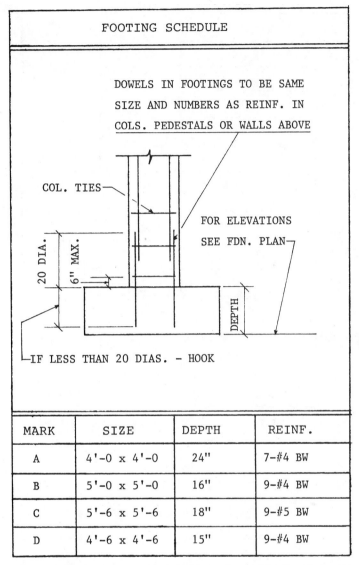

FOOTING SCHEDULE

DOWELS IN FOOTINGS TO BE SAME
SIZE AND NUMBERS AS REINF. IN
COLS. PEDESTALS OR WALLS ABOVE

MARK	SIZE	DEPTH	REINF.
A	4'-0 x 4'-0	24"	7-#4 BW
B	5'-0 x 5'-0	16"	9-#4 BW
C	5'-6 x 5'-6	18"	9-#5 BW
D	4'-6 x 4'-6	15"	9-#4 BW

Figure 12-11. Footing schedule includes a typical detail drawing and footing dimensions with identifying "Mark" for reference to plan drawing.

CONCRETE SLAB SCHEDULE				
SLAB MARK	BOTTOM RODS	TOP RODS	TEMP. STEEL	REMARKS
S6A	#4 @ 8	#5 @ 8	#4 @ 12	
S6B	#4 @ 6	#5 @ 7	#4 @ 12	
S6C	#4 @ 6	#5 @ 6	#4 @ 12	

(NUMBER IN SLAB COLUMN INDICATES SLAB THICKNESS

Figure 12-12. Concrete slab information is best described on a schedule. Slab Mark—"S6A" means Slab is 6 inches thick. Typical slab detail indicates location of bottom and top rods and the schedule indicates size.

RETAINING WALL SCHEDULE				
MARK		WALL A	WALL B	WALL C
D I M E N S I O N S	A	1'-2"	1'-2"	1'-3"
	B	5'-4"	4'-4"	3'-6"
	C	1'-4"	1'-1"	1'-0"
	D	2'-10"	2'-1"	1'-2"
	E	4'-6"	4'-2"	3'-9"
R E I N F O R C E M E N T	a	#5 @ 8	#4 @ 8	#4 @ 12
	b	#4 @ 8	#4 @ 12	#4 @ 12
	c	#4 @ 12	#4 @ 12	#4 @ 12
	d	#5 @ 10	#4 @ 10	#4 @ 12
	e	#4 @ 10	#4 @ 8	#4 @ 12

Figure 12-13. Some projects may require several different sized retaining walls and these should be listed and described in a schedule.

PEDESTAL SCHEDULE			
MARK	SIZE	VERTICAL REINF.	REMARKS
1	14 x 14	4-#3	-
2	13 x 13	4-#8	SEE DETAIL
3	18 x 20	8-#10	SEE DETAIL
4	18 x 18	8-#10	SEE DETAIL

ALL PEDESTAL TIES TO BE #3 @ 12"o/c AND IN
ACCORDANCE WITH A. C. I. DETAILING MANUAL

Figure 12-14. Pedestal schedules are frequently used on structural drawings to provide the necessary information and where required, reference is made to "See Detail".

ifications is written in note format on the drawings. On large, more complex projects, the general conditions and the technical section of the specifications may require several hundred pages to define the project fully.

Figure 12-10 to 12-14 show examples of structural schedules.

Appendix A

SPECIFICATIONS

Construction specifications are as much a part of contract documents as are the drawings (Figure A-1). One does not exist without the other. Contract documents consist of the Agreement, Conditions of the Contract, Drawings, and Specifications. On small simple projects, the entire specifications may be included on the drawings. More commonly, however, specifications are a separate entity, usually in the form of a manual. In fact, the American Institute of Architects (AIA) originated the term Project Manual and in 1964 adopted the concept and title ''Project Manual'' in lieu of the prior common title ''Specifications.''

The Construction Specifications Institute's MASTERFORMAT (MP-2-1) evolved from the format for the *AIA Project Manual* and consists of two broad categories of *documents* and *specifications.* The frontmatter of the MASTERFORMAT consists of the following documents:

> 00010 Pre-bid Information
> 00100 Instructions to Bidders
> 00200 Information Available to Bidders
> 00300 Bid Forms
> 00400 Supplements to Bid Forms
> 00500 Agreement Forms
> 00600 Bonds and Certificates
> 00700 General Conditions
> 00800 Supplementary Conditions
> 00850 Drawings and Schedules
> 00900 Addenda and Modifications

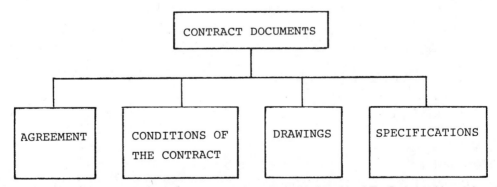

Figure A-1. Drawings are only part of the contract documents developed by A/Es. Both graphics and words are indispensable in describing projects. While drawings are essential to describe a construction project, specifications are equally essential to express requirements which can best be conveyed in words. Drawings should show the *form* of construction; specifications should establish its *quality*.

Drafters and designers are almost never involved with the frontmatter. These documents are developed by the project owners and their legal counsel and insurance advisors generally in consultation with the specifications writers in conjunction with architects and engineers as may be the case. But the technical portion of the specifications does have an impact on the work of drafters and designers.

MASTERFORMAT Specifications consist of 16 Divisions, with appropriate sections that cover all elements of the work as listed below (see also Figure A-2):

Division 1 General Requirements
Division 2 Sitework
Division 3 Concrete
Division 4 Masonry
Division 5 Metals
Division 6 Wood and Plastic
Division 7 Thermal and Moisture Protection
Division 8 Doors and Windows
Division 9 Finishes
Division 10 Specialties
Division 11 Equipment
Division 12 Furnishing
Division 13 Special Construction
Division 14 Conveying Systems
Division 15 Mechanical
Division 16 Electrical

According to CSI,

Specifications contain qualitative requirements for products, materials, and workmanship. Specifications must precisely define these qualities to assure use of correct materials and methods of

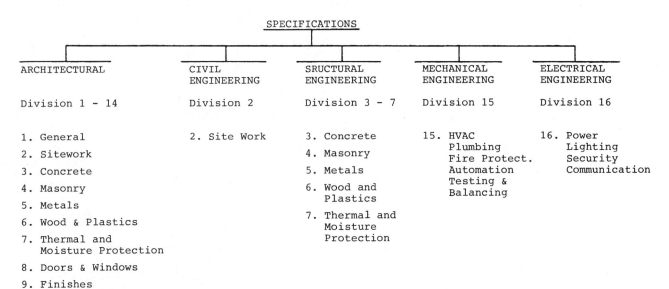

Figure A-2. Generally accepted division of responsibilities for major A/E disciplines. Some firms may have additional subdivisions within these major divisions.

assembly. The MASTERFORMAT provides a standard 16-division framework for organizing construction specifications for any project. . . .

Some A/E firms that develop their own specifications format use additional divisions. For example HVAC testing, adjusting, and balancing, when bid separately is identified separately as Division 17. In some instances, HVAC Automation is also identified as a separate division.

Drafters and designers are not involved with Division 1—General Requirements; Divisions 2–16, the technical part of the specifications, have the greatest impact on drawings and, therefore, the work performed by drafters and designers.

The following includes CSI Broadscope section title of Division 1. There are also Broadscope section title for all other Divisions.

CSI MASTERFORMAT—Division 1—General Requirements.

> 01010 Summary of Work
> 01020 Allowances
> 01025 Measurement and Payment
> 01030 Alternates
> 01040 Coordination
> 01050 Field Engineering
> 01060 Regulatory Requirements
> 01070 Abbreviations and Symbols
> 01080 Identified Systems
> 01090 Reference Standards
> 01100 Special Project Procedures
> 01200 Project Meetings
> 01300 Submittals
> 01400 Quality Control
> 01500 Construction Facilities and Temporary Controls
> 01600 Material and Equipment
> 01650 Starting of Systems/Commissioning
> 01700 Contract Closeout
> 01800 Maintenance

According to CSI—*Drawings are a graphic representation of the construction project; size and shape, general indication of materials and their location, connections and details, and diagrams, and isometrics depicting items such as mechanical and electrical systems. Schedules of structural elements, equipment, and finishes are also part of the drawings.*

Life would indeed be simple *if,* and that is a big *if,* drawings could be distinctly separate from specifications, and vice versa, i. e., if a drawing could illustrate construction and specifications could define construction. Unfortunately, this is not the case. Once writing appears on drawings, it must be coordinated with the specifications; however, specifications must be coordinated with more than just the writing on the drawings, specifications must be coordinated with the details and graphic designations of materials, systems, dimensions, details, etc., on the drawings. As a fundamental example, drawings contain notes, schedules and symbol lists, and these consist of terms that describe items of construction, and specifications must also address these.

The terminology must be identical in both cases, i. e., the "janitor's sink" noted on the drawings should not become "service sink" in the specifications.

Drafters and designers make the drawings, develop symbol lists, and provide schedules that are filled in at the appropriate time by qualified personnel. Specification writers develop the specifications. How the efforts of these two are coordinated depends to a great extent upon the firm or company for which one works. Some firms follow accepted practice, national standards, or their own specifically developed symbol list and equipment nomenclature. At the other extreme is the small firm or company that utilizes no standard and relies completely upon the *lead* person in the department to provide leadership, uniformity, and coordination.

As soon as possible after joining a firm or company, top priority for a drafter and designer is to investigate the existence of any and all standards and to obtain copies. Then faithfully comply with these standards unless of course there are errors or omissions, then corrective action should be taken, but not unilaterally.

Why so much stress on specifications and uniformity between drawings and specifications in a book on drafting? First, confusion created by inconsistency between drawings and specifications can cause bidding errors and second, can result in misinterpretation during construction, which could cause delays and/or costly overruns.

Both graphics and words are necessary in describing construction projects. Drawings should graphically depict the form of construction; specifications should establish its quality. Unfortunately, drawings also contain words in the form of notes, symbols, and schedules, and these words must not conflict with the words in the specifications. Drawings and specifications serve complimentary functions. Specifications should supplement, and not repeat, information shown on drawings.

It is for these reasons that a chapter on specifications is included in this book on drafting and design.

1. RELATIONSHIP BETWEEN DRAWINGS AND SPECIFICATIONS

There is a very distinct relationship between what is written on drawings and what is included in specifications. As stated above, drawings usually contain notes, symbols, and schedules. Of the three, notes generally tend to create the greatest confusion for the inexperienced drafter.

Notes are used as a means of communications and are for either clarification or instruction. Specification information should never be included in notes. On small projects, where the specifications are placed on the drawings, the specifications should be under the heading of *Specifications* and notes under the heading of *Notes.* The separation must be as distinct in this case as where the specifications are a free-standing document.

Notes may be either on the body of the drawing with leaders indicating the point of application or grouped under the heading *Notes* along the right side of the drawing. Notes should be kept to a minimum so as not to clutter the drawing unnecessarily. Some notes are circled-up with a dark heavy line for special attention.

Clarification notes are frequently used in lieu of large-scale detail drawings. Sometimes a few words are sufficient to clarify the intent; in those cases, the few words are preferred because the note can be placed at the point of application and eliminates the need for looking for a detail that may be on another drawing somewhere in the set. Also, making a note usually consumes less time and remember, *time is money.*

Notes of instruction are for the contractor; they tell the contractor what to do or not to do. A very good example is Figure A-3, Addition of Cabinet Unit Heater in Stair No. 3. This drawing was issued as a Bulletin and part of an existing construction contract. Notice that there is no information on the type of pipe, insulation, valve and low pressure steam trap assembly as this information is covered in the specifications. If this were part of the construction drawings, information on the unit heater would not be on the drawings, at least not in this location. That information could be either in the Unit Heater Schedule or inserted into the specifications, depending upon office practice.

Normally, trade names should be avoided on drawings, all reference to equipment and materials should remain generic. When proprietary products are required, these should be covered in the specifications. The primary advantage for the exclusion of proprietary information on drawings is in the event of an acceptance of a substitution. In this case, drawing revisions will not be required. All the changes will be made, if required, in the specifications.

However, Figure A-3 does contain specific proprietary information. This is a free

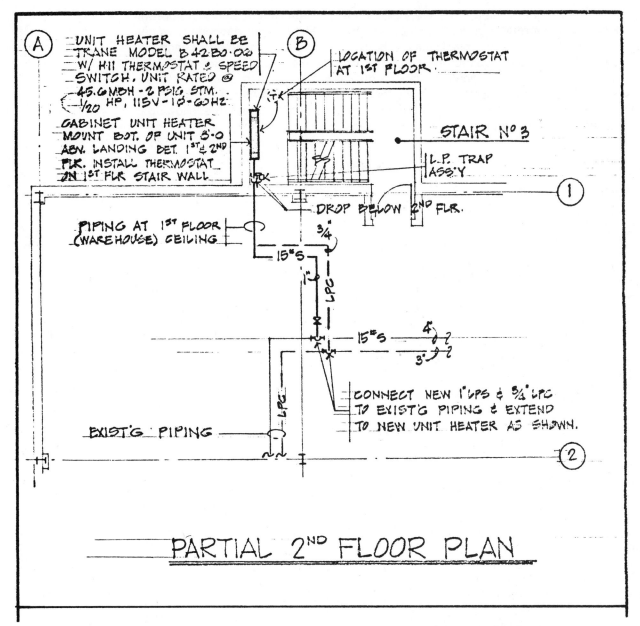

Figure A-3. Partial 2nd floor plan issued as an addition to an existing construction contract. The original specifications describe methods of installation and materials of construction. This drawing describes the addition of a unit heater in Stair No. 3.

standing document and was issued to the contractor for inclusion in an existing construction project and no additional information was required in the specifications. While it is standard practice to exclude trade names from drawings, there are exceptions.

As the entry level drafter gains experience and is given more latitude in the decision making process, the relationship between drawings and specifications must be fully understood. This relationship is best explained by including several typical specification sections to point out the type of information included in the specifications and what need not be addressed on the drawings.

2. TYPICAL SPECIFICATIONS

Since not all A/E firms use CSI's MASTERFORMAT specifications, the typical specifications included are intended to show what is covered in them for information for drafters and designers and are not intended for specification writers, and therefore will not follow Masterformat system. Drawings should identify, but not describe material, components, or equipment. Specifications should describe in complete detail each material or equipment and required quality, and explain the method of installation.

The front section of specifications lists all drawings that are part of contract documents. The essential consideration here is that the drawing title match the drawing list. While there is no hard and fast rule that determines title nomenclature, the following is an example of a set of drawings:

DRAWING LIST

Drawing No.	Title
A-1	Cover Sheet
A-2	Site Plan
A-3	First Floor Plan
A-4	Second Floor Plan
A-5	Reflected Ceiling Plan
A-6	Roof Plan
A-7	Elevations and Sections
A-8	Wall Sections and Details
A-9	Details and Schedules
S-1	Foundation Plan
S-2	Second Floor Plan
S-3	Roof Plan
S-4	Column and Beam Schedules
S-5	Details
P-FP-E-1	Site Plan, Plumbing, Fire Protection and Electrical Plan
P-2	First Floor Plan
P-3	Second Floor Plan
P-4	Roof Plan
P-5	Riser Diagrams and Details
H-1	First Floor Plan
H-2	Second Floor Plan
H-3	Roof Plan
H-4	Schedues and Details
FP-1	First Floor Plan
FP-2	Second Floor Plan
FP-3	Schedules and Details
E-1	First Floor Lighting Plan
E-2	Second Floor Lighting Plan
E-3	First Floor Power Plan
E-4	Second Floor Power Plan
E-5	Single-line Diagram, Schedules and Details

Letter designations are rather straightforward, but since additional ones are sometimes used, the following list defines their use:

A = Architectural

S = Structural

P = Plumbing

H = HVAC

FP = Fire Protection

E = Electrical, may include I = Instrumentations

*M = Mechanical, may include mechanical, HVAC, plumbing and automation

*Some firms use M for boiler plants, steam systems, etc.

Small projects frequently use four designations: architectural, structural, mechanical, which includes HVAC, plumbing, fire protection, and automation, and electrical. Sometimes this is reduced to three; architectural, which includes civil and structural, mechanical, and electrical. The primary significance of designation is for identification, and on larger projects for separation by trades.

Drawing titles should be as simple as possible, yet descriptive. The letter designation identifies the architectural and engineering disciplines. When several disciplines are included on one drawing, the letter designation of each discipline should appear in the title block, for example, when a site plan includes plumbing, electrical, and fire protection work, that drawing should be identified as: P-E-FP-1. Generally, only site plans are combined; others are usually by specific discipline. Avoid redundancy, such as Drawing No. E—1, Electrical Floor Plan. E—1 identifies the drawing as electrical, so the word description is unnecessary.

a. Architectural

Architectural work is described in several divisions of the specifications, Divisions 2 through 14. Civil engineering is included in Division 2, and some civil engineering work may be found in Divisions 11 and 13. Structural work is described in Division 3, 4, 5, 6 and 7. Division 15 covers mechanical, HVAC, plumbing, fire protection, automation, and testing and balancing. Division 16 covers electrical power and light, security, communications, and instrumentation.

One of the major problems encountered in developing contract documents is the lack of thorough *coordination* of drawings with specifications. The second area of concern is the inclusion of correct information in the wrong division and section of the specification. This is where the CSI MASTERFORMAT provides great service. Its divisions are subdivided into numerous sections covering most of the usual subjects encountered in building construction.

The following typical specifications are included to show what is in specifications and what is intended to be shown on drawings. There are several statements in specifications for reference to information on drawings, "shown in detail," "where indicated on the drawings," "on the Finished Schedule," "Fixture Schedule on the drawings," to name a few. Review the typical specifications very carefully to note the general and specific language used.

Typical Specification

Concrete-Slab and Finished Floor on Ground

Concrete slabs shall be reinforced with 6 in. × 6 in. × 8/8 mesh lapped not less than 6 in. which shall be placed 1 in. below the finished floor. In addition the slab shall be reinforced at exterior walls as shown in detail and the partition footings shall have reinforcement as shown. Partition reinforcement shall be centered on the partitions and shall be placed 1 in. above the bottom of the footings as shown in detail.

First, this section of the specification covers reinforcement, eliminating the need for that information on the drawings. Second, the second sentence states, . . . shall have reinforcement as shown. That means on the drawings and it must be there.

This simple short paragraph shows why drawings must be coordinated with the specifications. Another excellent example is the following:

Wall or ceiling access panels where indicated on the drawings shall be furnished and delivered to the site for erection by others.

(1) Wall and ceiling access panels in plastered areas shall be equal to Hohmann and Barnerd, Inc., Type #701-E, in masonry walls Type #702. In plastered areas same to be delivered to lather for setting, in masonry walls to masonry Contractor for building in.

(2) Access panels shall be of sizes indicated on drawings. Where specific size is not indicated access panels shall be of 24″ × 30″.

Please note "where indicated on the drawings," and under (1), the trade name and type are listed. Access panels and locations must be shown on the drawings; the make and type are not required.

The section under Resilient Flooring states:

The work under this section shall consist of the furnishing of labor, materials and equipment required for the proper installation of resilient flooring, base, and related work in areas called for on the Finish Schedule and called for herein.

Then goes on to describe, by trade name, vinyl tile, sheet vinyl, vinyl cove base, and edge strips. The specifications also address workmanship. Please note in the above section "as called for on the Finish Schedule." That information is on the drawings.

This same philosophy continues through the entire specification and for all trades. It is essential that all but the lowest level-entry drafter know what to show on and what to omit from the drawings.

b. Structural

Structural work is included under the CSI MASTERFORMAT Division 3—Concrete, Division 5—Metals and Division 7—Wood and Plastic. The specifications address such items as *Scope of Work, Standard Codes, Materials, Framing Connections, Bearing Plates, Erection, Painting,* and *Workmanship.* Again in this work there will be such statements as, *as indicated on the drawings,* and *on schedules.*

c. Mechanical

Division 15 covers all work as described earlier and relates to work done by the mechanical trades. On large projects, this work may be divided between several subcontractors. One essential statement generally found in this division is as follows:

This specification and the accompanying drawings are intended to illustrate and include all labor, material and equipment required for the Mechanical Work. . . .

A typical section under Motors includes the following:

Motors shall be ball bearing unless otherwise noted. Ball bearing motors shall be equipped with lubricating type bearings. Each bearing housing shall have one grease fitting and one removable plug in the bottom of the grease sump to provide for flushing and pressure relief when lubricating. Motors shall be permanently marked that bearings are lubricating type.

Motors ½ HP and larger shall be 480 volt, 60 cycle, 3 phase. Motors under ½ HP shall be 120 volts, 60 cycle, single phase. All motors shall be rated 40°C., NEMA rating, constant speed, squirrel cage induction type. Drip-proof type shall be the minimum requirement, with other types provided where noted or required.

Single-phase motors shall be capacitor start type. Two speed, 3-phase motors shall be the two-winding type and two speed starters will be furnished by the Electrical Contractor.

Since this section describes the type and rating, etc., of the motors required, it is redundant to include that information on the drawings.

d. Electrical

The following section under Scope of Work is relevant to the issue at hand, coordination. In this case, in addition to coordination between the drawings and the specifications, there also needs to be coordination between the electrical and mechanical drawings. The following are pertinent sections from Division 16—Electrical:

Electrical motors and pilot control devices in connection with the building services equipment provided in other contracts are included in such other contracts unless specified herein. However, the furnishing and installation of the motor control equipment is included in the electrical construction contract, together with the furnishing of any mounting supports necessary.

Electric wiring to the motor control apparatus and from the control apparatus to the motors, equipment and building services equipment, including the final connections to such motors and apparatus, and the adjustment of connections for proper rotation, are included in the electrical contract.

The third paragraph of the Section—Lighting Fixtures, General is important for drafters and designer to understand. See below:

1-a This Specification covers the furnishing and installation of electrical lighting fixtures in the new building addition.

1-b Lighting fixtures and auxiliaries shall be manufactured in accordance with standards and requirements of the Underwriters' Laboratories, Inc., and shall bear their label of approval.

1-c Fixtures shall be as specified in detail in the "Fixture Schedule" on the drawings.

Another important coordination item between drawings and specifications is the "Outlet Mounting Height", see typical specification section below:

2-a Unless otherwise indicated on the drawings, outlets shall be located as follows:

1. Lighting Switches	4'-6"
2. Wall Receptacles	18"
3. Motor Controllers—General	4'-6"
4. Light and Power Panels	6'-6" to top, maximum

2-b Unless otherwise noted, all dimensions are to the center of the finished outlets with all apparatus in place.

This eliminates the need for indicating the outlet mounting height on the drawing, except in the case of special requirements for unique equipment.

From the above it should be rather obvious to drafters and designers what type of information is generally included in the specifications and why thorough coordination between the different disciplines involved in making the drawings, and the drawings and specifications, is absolutely necessary. Including specification information on drawings increases the probability of confusion between drawings and specifications, and also wastes valuable drafting time.

3. DRAWING NOTES

Drawing notes are an essential part of construction drawings; this is a fact of life, and as such, the subject must be addressed. Notes are included in the chapter on specifications because of the relationship between drawings and specifications.

Since what notes are needed has already been explained, this section will deal with where they are placed on drawings, and how to use them most effectively. Notes may be placed on the body of a drawing or in one location under the heading NOTES along the right-hand side of the sheet. Where the notes are placed on a drawing has a lot to do with when they are placed on the drawing.

When notes are placed in the body of a drawing the drawing must be virtually completed before any notes are placed on the drawing. For example, all hard lines and all dimensioning must be finished first; then, the notes can be neatly inserted in the remaining space. The drawing should never be completed after the notes.

When notes are placed in one common location along the side of a drawing, it generally makes no difference when they are inserted since this space is reserved for notes. The only essential consideration with this scheme is that the order of the notes must follow some definite plan. For example, notes dealing with specific equipment must be assembled into groups, and not interspersed throughout a long list of notes. This is sound practice, and it eliminates possible confusion, since the contractor will find all pertinent information relative to that piece of equipment in one location. The contractor or others reading the drawings can readily find all the necessary information in one location and need not be concerned about missing something.

Notes placed in one location should have a definite relationship with the drawing for simplification. Imagine reading a drawing like reading a book, from left to right and from top to bottom. The notes should be placed in that order.

This method is used when a large number of notes are required. Many notes unnecessarily clutter a drawing and can create confusion. A well constructed, neat drawing can be completely destroyed when all white space is totally filled with notes. A drawing should be *a graphic representation of a construction project.*

With this method a ten-word note is replaced by a simple *See Note 1* or *Note 1* adjacent to the point of application and a short leader indicating the exact location.

In cases where few notes are anticipated, these may be placed in the body of the drawing. Now the drawing must be complete and fully dimensioned. Notes should never interfere with line work and dimensioning. A long leader from the note to the point of application is preferred to crowding the note onto the drawing.

Finally, notes should be concise and to the point, never write more than is absolutely necessary. The letters should be neat and legible in order to avoid confusion. Letters should be made so that they are never misinterpreted for numerals.

Appendix B

CONVERSION TABLES

Conversion tables provide a simple means for converting units from one to another. Table B. 1 contains those units commonly used in building construction drafting and design. To use the table, multiply the first unit by the conversion factor to obtain the answer.

EXAMPLE: 2 acres × 43,560 = 87,120 square feet.

Table B.1

MULTIPLY	BY	TO OBTAIN
acres	43,560	square feet
atmospheres (U. S. Standard)	29.92	inches of mercury
atmospheres (U. S. Standard)	33.90	feet of water
atmospheres (U. S. Standard)	14.696	pounds per square inch
atmospheres (U. S. Standard)	101.325	kilo-pascals
British thermal units	777.5	foot-pounds
British thermal units	2.928×10^{-4}	kilowatt-hours
Btu per minute	0.02356	horsepower
Btu per minute	0.01757	kilowatts
Btu per minute	17.5798	watts
centimeters	0.3937	inches
cord	64	cubic feet
Cubic feet	62.43	pounds of water
Cubic feet	1728	cubic inches
Cubic feet	0.037	cubic yards
Cubic feet	7.481	gallons
cubic feet per minute	0.12468	gallons per second
cubic feet per minute	62.4	pounds of water per minute
cubic feet per second	4489	gallons per minute
cubic inches	5.787×10^{-4}	cubic feet
cubic yards	27	cubic feet
cubic yards	46,656	cubic inches
cubic yards	202.0	gallons
degrees (angle)	60	minutes
degrees (angle)	3600	seconds
degrees (angle)	0.01745	radians
feet	30.48	centimeters
feet	12	inches
feet	0.3048	meters
feet of water	0.0295	atmospheres
feet of water	0.8826	inches of mercury
feet of water	62.43	pounds per square foot
feet of water	0.4335	pounds per square inch
foot-pounds per minute	3.03×10^{-5}	horsepower
foot-pounds per minute	2.26×10^{-3}	kilowatts
foot-pounds per second	1.818×10^{-3}	horsepower

Table B.1 (Continued)

MULTIPLY	BY	TO OBTAIN
furlong	40	rods
gallons	8.345	pounds of water (at 60°F)
gallons	0.1337	cubic feet
gallons	231	cubic inches
gallons	4.951×10^{-3}	cubic yards
gallons	3.785	liters
gallons	8	pints (liquid)
gallons	4	quarts (liquid)
gallons per minute	2.228×10^{-3}	cubic feet per second
horsepower	42.44	Btu per minute
horsepower	33,000	foot-pounds per minute
horsepower	550	foot-pounds per second
horsepower	0.746	kilowatts
horsepower	746	watts
Horsepower (boiler)	33,520	Btu per hour
Horsepower (boiler)	9.804	kilowatts
horsepower-hours	2547	Btu
horsepower-hours	0.746	kilowatt-hours
inches	2.54	centimeters
inches	0.001	mils
inches of mercury	0.03342	atmospheres
inches of mercury	1.133	feet of water
inches of mercury	70.73	pouunds per square foot
inches of mercury	0.4912	pounds per square inch
inches of water	0.0736	inches of mercury
inches of water	5.1948	pounds per square foot
inches of water	0.0361	pounds per square inches
kilograms	2.205	pounds
kilometers	3280.8	feet
kilometers	0.6214	miles
kilometers	1093.6	yards
kilowatts	56.92	Btu per minute
kilowatts	1.341	horse-power
kilowatt-hours	3413	British thermal units
kilowatt-hours	1.341	horse-power-hours
meters	3.2808	feet
meters	39.3701	inches
meters	1.0936	yards
miles (Land)	5280	feet
miles (Land)	1.609	kilometers
miles (Land)	1760	yards
miles per hour	88	feet per minute
miles per hour	1.609	kilometers per hour
ounces (weight)	0.0625	pounds
pints (liquid)	28.87	cubic inches
pounds	16	ounces (weight)
pounds of water	0.016	cubic feet
pounds of water	27.68	cubic inches
pounds of water	0.1198	gallons
pounds per square foot	0.0160	feet of water
pounds per square foot	6.94×10^{-3}	pounds per square inch
pounds per square inch	0.0680	atmospheres
pounds per square inch	2.31	feet of water
pounds per square inch	2.036	inches of mercury
pounds per square inch	144	pounds per square foot
quarts	32	fluid ounces
quarts (dry)	67.2	cubic inches
quarts (liquid)	57.8	cubic inches
rods	16.5	feet
square centimeters	0.155	square inches
square feet	2.296×10^{-5}	acres
square feet	144	square inches

Table B.1 (*Continued*)

MULTIPLY	BY	TO OBTAIN
square inches	6.4516	square centimeters
square inches	6.94×10^{-3}	square feet
square miles	640	arcres
square miles	2.590	square kilometers
square yards	0.8361	square meters
square yards	9	square feet
temp (degs C) + 17.778	1.8	temperature (°F)
temp (degs F) − 32	$5/9$	temperature (°C)
tons (long)	2240	pounds
tons (short)	2000	pounds
yards	.9144	meters

Table B.2. English/Metric conversion table for commonly used fractions in building construction drafting.

ENGLISH FRACTIONS (INCHES)	ENGLISH DECIMAL (INCHES)	METRIC (MILLIMETER)
$1/64$	0.015625	0.3969875
$1/32$	0.03125	0.79375
$1/16$	0.0625	1.5875
$1/8$	0.125	3.175
$1/4$	0.250	6.350
$3/8$	0.375	9.525
$1/2$	0.500	12.700
$5/8$	0.625	15.875
$3/4$	0.750	19.050
$1''$	1.000	25.400
$1 1/2$	1.500	38.100
$3''$	3.00	76.200

A/E firms serving the international market may be required to prepare drawings using the metric scale. Table B.3 provides metric scales that correspond (not equal to) typical English scale fractions used on building construction drawings:

Table B.3. Metric scale corresponding to English scales

ENGLISH SCALES FRACTIONS (INCHES)	METRIC SCALES (MILLIMETERS)
$1/16'' = 1'\text{-}0''$	5 mm = 1 m (1:200)
$1/8'' = 1'\text{-}0''$	10 mm = 1 m (1:100)
$1/4'' = 1'\text{-}0''$	20 mm = 1 m (1:50)
$3/8'' = 1'\text{-}0''$	30 mm = 1 m (1:333)
$1/2'' = 1'\text{-}0''$	40 mm = 1 m (1:25)
$3/4'' = 1'\text{-}0''$	65 mm = 1 m (1:15.385)
$1'' = 1'\text{-}0''$	100 mm = 1 m (1:10)
$1 1/2'' = 1'\text{-}0''$	125 mm = 1 m (1:8)
$3'' = 1'\text{-}0''$	250 mm = 1 m (1:4)

Appendix C

ABBREVIATIONS

An abbreviation is a shortened form of a written word or phrase that is used in place of the whole. The use of the appropriate abbreviation on building construction drawings saves a significant amount of time. Additionally, abbreviations, properly used, aid in freeing drawings from unnecessary clutter, and since drawings are graphical presentations, clutter detracts from the illustration intended. Also, why print Air Conditioning Unit No. 1 when AC-1 will suffice.

Not every abbreviation has the same meaning in the different disciplines, for example: AD on a plumbing drawing may be used to represent *AREA DRAIN,* whereas on the HVAC drawing it may represent *ACCESS DOOR.* While this may be obvious to most contractors, it is good practice to include a list of abbreviations on every set of discipline drawings to show what is intended.

Abbreviations should not be used indiscriminately. When some word or phrase occurs only once, it may be better to spell out the word completely. Since many companies and A/E firms have a list of standard abbreviations, that list must be used on all drawings developed by that office. Abbreviations generally mean singular or plural, as may be the case.

The following is a list of some of the more common abbreviations used on building construction drawings.

1. ARCHITECTURAL AND SITE WORK

AFF	above finished floor
AP	access panel
AT	asphalt tile
BM	bench mark
BZ	bronze
BZAL	bronze finish aluminum
CAB	cabinet
CAR	carpet
CB	catch basin
CL	closet
CLG	ceiling
CMU	concrete masonry unit
CO	clean out
CON	concrete
CT	ceramic tile
CU	copper
DET	detail
DIM	dimension
DN	down
DR	drain

EP	epoxy paint
FH	fire hydrant
FT	foot
GR	grade
GWB	gypsum wall board
HB	hose bib
IN	inch
KO	knock out
MH	manhole
NIC	not in contract
NTS	not to scale
OC	on center
P or PTD	painted
PL	property line
PN	panel
QT	quarry tile
RD	roof drain
RWI	rain water inlet
RWC	rain water conductor
S	sanitary sewer
SCHD	schedule
SECT	section
SS	storm sewer (on site plan)
SS	service sink (on floor plan)
T	tread
TBR	to be removed
TER	terrazzo
TYP	typical
V	vinyl
VAT	vinyl asphalt tile
VCB	vinyl cove base
WC	water closet
WI	wrought iron
WV	water valve

2. HVAC (MECHANICAL)

AD	access door
A/C	air conditioner
AP	access panel
ATC	automatic temperature control
BDD	back draft damper
CD	ceiling diffuser
CENT	centrifugal
CFM	cubic feet per minute
DV	drain valve
EF	exhaust fan
EP	electric-pneumatic relay (solenoid air valve)
EXH	exhaust
F	fan
FAI	fresh air intake
FD	fire damper
FPM	feet per minute
FS	fire stat

F & T	float and thermostatic trap
FZ	freeze stat
GD	gravity damper
GPM	gallons per minute
H	humidistat
HOA	hand-off-automatic
HP	horsepower
LD	linear diffuser
MBH	thousand Btu per hour
MH	manhole
NC	normally closed
NIC	not in contract
NO	normally open
OA	outside air
PE	pneumatic-electric relay
PRV	pressure reducing valve
PSF	pounds per square foot
PSI	pounds per square inch
RA	return air
RF	roof fan
SA	supply air
SD	smoke damper
T	thermostat
UH	unit heater
V	vent
VD	volume damper

3. PLUMBING AND FIRE PROTECTION

AD	area drain
AP	access panel
CB	catch basin
CI	cast iron
CO	clean out
CW	domestic cold water
DF	drinking fountain
DSP	dry stand pipe
DV	drain valve
EWC	electric water cooler
F	fire line (on site plan)
FEC	fire extinguishing cabinet
FH	fire hydrant
FHC	fire hose cabinet
FHR	fire hose rack
HB	hose bib
HW	domestic hot water
INV	invert
LAV	lavatory
MH	manhole
MR	mop receptor
NIC	not in contract
NTS	not to scale
OWH	outside wall hydrant
PRV	pressure reducing valve

PSI	pounds per square inch
RD	roof drain
RWC	rain water conductor
SHR	shower
SS	service sink
SWV	soil, waste and vent
UR	urinal
V	vent
VTR	vent through roof
WC	water closet
WF	wash fountain
WH	wall hydrant
WSP	wet stand pipe

4. ELECTRICAL

AFF	above finished floor
AS	automatic starter
CB	circuit breaker
CKT	circuit
DISC	disconnect
EP	emergency panel
EWC	electric water cooler
HOA	hand-off-automatic
HP	horsepower
HR	home run
KVA	kilovolt ampere
KW	kilowatt
LP	lighting panel
MCC	motor control panel
MS	manual starter
NIC	not in contract
NTS	not to scale
PA	public address (paging system)
PB	push button station
PE	pneumatic-electric relay
PF	power factor
PN	panel
PL	pilot light
S	single-pole switch
S_2	double-pole switch
S_3	three-way light switch
S_4	four-way light switch
S_d	dimmer switch
S_k	key operated switch
SW	switch
WP	weather proof

Appendix D

ENGINEERING FORMULAS

Engineering drafters seeking to develop their skills and abilities to advance to designer status must become familiar with engineering formulas used in the different disciplines. The number of formulas used in developing building construction drawings is rather limited, but occur repeatedly in project after project. These formulas can be divided into two broad categories; electrical and mechanical, with some overlap between the two disciplines. For example, mechanical engineers use motors to drive fans, pumps, and compressors, and electrical engineers provide the circuitry to power the motors.

What follows is a list of the more commonly used formulas. Definitions, where appropriate, are included. Equation derivations are not attempted, but certain constants, where necessary, are explained.

1. ELECTRICAL FORMULAS

Ohm's Law is fundamental to electrical formulas, and states that the intensity of current (amperes) in any circuit is equal to the electromotive force (volts) divided by the resistance of the circuit.

$$\text{Amperes} = \text{Volts} / \text{Ohms}$$
$$\text{Volts} = \text{Amperes} \times \text{Ohms}$$
$$\text{Ohms} = \text{Volts} / \text{Amperes}$$

a. Direct Current Circuits

The unit of electrical power is the watt or kilowatt (1000 watts) and is equal to volts multiplied by amperes.

$$\text{Watts} = \text{Volts} \times \text{Amperes}$$
$$= \text{Ohms} \times \text{Amperes} \times \text{Amperes}$$
$$= \text{Ohms} \times \text{Amperes}^2$$
$$\text{Amperes} = \text{Watts} / \text{Volts}$$
$$\text{Kilowatts (kw)} = \text{Volts} / \text{Amperes} / 1000$$
$$\text{Kilowatt-Hours} = \text{Volts} \times \text{Amperes} \times \text{Hours} / 1000$$
$$\text{Horsepower} = \text{Volts} \times \text{Amperes} \times \text{Efficiency} / 746$$

A motor's nominal horsepower rating refers to the mechanical power available at the motor shaft at full load. Motors with a "continuous" rating deliver the rated horsepower of the motor without exceeding the temperature rise.

The service factor of a motor is the maximum overload that can be applied to the motor without exceeding the temperature limitations of the motor insulation when voltage and frequency are maintained at name plate values and the ambient temperature does not exceed 40°C (104°F). The motor may be loaded up to the horsepower obtained by multiplying the rated horsepower by the service factor.

$$\text{Motor Efficiency} = 100 \times \text{Output} / \text{Input}$$
$$\text{Motor Output} = \text{HP} \times 746$$
$$\text{Motor Input} = \text{HP} \times 746 / \text{Efficiency}$$

b. Alternating Current Circuits

Single Phase Circuits

$$\text{Watts (W)} = \text{Volts (V)} \times \text{Amperes (A)} \times \text{Power Factor (PF)}$$
$$\text{KW} = \text{V} \times \text{A} \times \text{PF} / 1000$$
$$= \text{Kilovolt-Amperes (KVA)} \times \text{PF}$$
$$\text{PF} = \text{Watts} / \text{Volt-Amperes}$$
$$= \text{KW} / \text{KVA}$$
$$= \text{Cosine } \theta$$
$$= \text{Actual Power (KW)} / \text{Apparent Power (KVA)}$$
$$\text{Horsepower} = \text{V} \times \text{A} \times \text{PF} \times \text{Efficiency} / 746$$

Three Phase Circuits

$$\text{KW} = 1.732 \times \text{V} \times \text{A} \times \text{PF} / 1000$$
$$\text{HP} = 1.732 \times \text{V} \times \text{A} \times \text{PF} \times \text{Efficiency} / 746$$

c. Motor Application

$$\text{Work (ft-lbs)} = \text{Force} \times \text{Distance}$$
$$\text{Power} = \text{Work} / \text{Time} = \text{ft-lbs per second (or minute)}$$
$$\text{1 Mechanical HP} = 33,000 \text{ ft-lbs per minute}$$
$$= 550 \text{ ft-lbs per second}$$
$$= 746 \text{ Watts}$$
$$= 2545 \text{ Btu per hour}$$
$$= 42.43 \text{ Btu per minute}$$
$$\text{Torque (ft-lbs)} = \text{HP} \times 5250 / \text{RPM (Full Load)}$$

For Pumps

$$\text{Brake Horsepower (BHP)} = \text{GPM} \times \text{H} \times \text{SG} / 3960 \times \text{PME}$$
$$= \text{GPM} \times \text{P} \times \text{SG} / 1714 \times \text{PME}$$

where : GPM = Gallons per minute
       H = Dynamic head in feet
       P = Pressure in pounds per square inch
     SG = Specific gravity (1 for water)
   PME = Pump mechanical efficiency

For Fans and Blowers

$$\text{Air HP} = \text{Q} \times \text{H} / 6350$$
$$\text{BHP} = \text{Q} \times \text{H} / 6350 \times \text{FME}$$

where: Q = Air quantity, in cubic feet per minute
     H = Head, inches of water gauge
   FME = Fan mechanical efficiency

d. Electric Heating

Electricity is sometimes used to heat water and air in buildings. The amount of heat required to raise the temperature of a substance is:

$$H = C \times W \times TD$$

where: H = BTU
      C = Specific heat of the substance
      W = Weight of the substance in pounds
    TD = Temperature difference

A more practical form of this formula is one where the answer is in watts or kilowatts, and the substance is in gallons per unit of time, or cubic feet per minute. For water problems the specific heat is omitted since C = 1.

$$Watts = 2.44 \times GPH \times TD$$
$$146.5 \times GPM \times TD$$

where: GPH = Gallons per hour
    GPM = Gallons per minute

For air problems the specific heat is C = 0.24.

$$Watts = 0.316 \times CFM \times TD$$

where: CFM = Cubic feet per minute

2. MECHANICAL FORMULAS

The list of mechanical formulas included here is a bit more extensive than that for the electrical discipline. Basically, mechanical engineering uses the properties of air and water in solving HVAC problems. For example, air is ducted from one point to another, similarly, water is piped from one location to another, and both are either heated or cooled, as necessary. These problems usually involve temperatures and volumes of the air and water, and fans and pumps are used to move the fluid. While there are various gases and liquids that may be encountered in mechanical engineering, air and water are the most common in HVAC.

a. Air

Air is measured in either degrees Celsius (formerly Centigrade) (°C) or degrees Fahrenheit (°F).

$$°C = 5/9 \times (°F - 32)$$

$$°F = 9/5 \times °C + 32$$

Standard air is air with a density of 0.075 lb per cu ft and an absolute viscosity of 1.225×10^{-5} lb (mass) per ft-sec. It is substantially equivalent to dry air at a temperature of 70°F and atmospheric pressure of 14.7 psia. At these conditions one pound of air occupies a volume of 13.34 cu ft. Manufacturers of fans and other HVAC equipment rate equipment at standard conditions.

The British thermal unit (Btu) is the accepted measurement of heat, one Btu is the amount of heat required to raise the temperature of one pound of water one degree Fahrenheit. The term MBH is used to express 1000 Btu.

The amount of heat required to raise the temperature of a substance is:

$$H = C \times W \times TD$$

where: H = Heat, Btu
     W = Weight of the substance (pounds)
    TD = Temperature change of the substances (°F)
     C = Specific heat of the substance (Btu per lb-°F)
     C = For water, 1.000 Btu/lb-°F
     C = For air, 0.24 Btu/lb-°F

The specific heat of air at normal HVAC temperatures is 0.2392 Btu per lb-°F and is rounded off to 0.24. There are three formulas commonly associated with HVAC systems: (air side)

$$\text{Sensible Heat (SH)} = 1.08 \times CFM \times TD$$
$$\text{Latent Heat (LH)} = 0.68 \times CFM \times GD/\text{lb dry air}$$
$$\text{Total Heat (TH)} = 4.5 \times CFM \times HD$$

where: TD = °F temperature difference, dry bulb
    GD = Grains per pound of air, difference
    HD = Change in heat content in Btu per lb

Fans are used to move air in HVAC systems:

$$\text{Fan Horsepower} = Q \times H / 6350$$
$$\text{Brake Horsepower} = Q \times H / 6350 \times FME$$

where: Q = Air Quantity, cfm
     H = Water gauge pressure in inches
  FME = Fan mechanical efficiency

The performance of the same fan for different speeds and for same air densities, and the performance of similar fans operating at the same tip speed and for the different densities can be predicted by the fan laws.

For the same fan handling the same gas (air) at different speeds use the following formulas:

• Quantity (CFM) varies directly with fan speed:

$$Q_2 = (N_2/N_1) \times Q_1$$

• Head (static pressure) varies as the speed squared:

$$H_2 = (N_2/N_1)^2 \times H_1$$

• Horsepower varies as the speed cubed:

$$HP_2 = (N_2/N_1)^3 \times HP_1$$

• Amperage varies as the speed cubed:

$$A_2 = (N_2/N_1)^3 \times A_1$$

The number of air changes within a space is as follows:

$$\text{Number of air changes (N)} = 60 \times \text{CFM} / \text{Cu Ft of space}$$
$$\text{Volume of air (CFM)} = \text{N} \times \text{Cu Ft} / 60$$

where: Q_2 = Final volume, in cubic feet per minute
Q_1 = Initial volume, in cubic feet per minute
N_2 = Final speed in revolutions per minute
N_1 = Initial speed in revolutions per minute
H_2 = Final head in pounds per square inch
H_1 = Initial head in pounds per square inch
HP_2 = Final horsepower
HP_1 = Initial horsepower
A_2 = Final current amperes
A_1 = Initial current in amperes

The volume of air at any other temperature can be determined from: $P_2 \times V_2 / T_2 = P_1 \times V_1 / T_1$,

$$\text{Degrees Ranking} = °F + 459.678 \text{ (Rounded off to 460)}$$

when $P_2 = P_1$; (constant pressure)

$$V_2 = V_1 \times 13.34 \times (460 + T) / (460 + 70)$$
$$= V_1 \times (460 + T) / 39.7$$

where V = Volume of air in cubic feet at temperature T (°F)
T = Dry bulb temperature (°F)

The actual volume of air in cubic feet per minute at any other temperature different from standard cubic feet per minute is:

$$\text{ACFM} = \text{SCFM} \times (460 + T) / 530$$
$$\text{SCFM} = \text{ACFM} \times 530 / (460 + T)$$

where: 530 = 460°F + 70°F (Standard temperature)

The diameter of the driver and driven pulleys for various shaft speeds can be determined from the following relationships:

$$d \times n = D \times N$$

where: d = Diameter of driver pulley
n = Speed of the driver pulley (rpm)
D = Diameter of the driven pulley
N = Speed of the driven pulley (rpm)
RPM = Revolutions per minute

also:

$$N = d \times n / D$$
$$D = d \times n / N$$
$$n = D \times N / d$$
$$d = D \times N / n$$

b. Water

The density of a substance is mass per unit volume. A gallon of water is found to have a mass of 8.345 lbs at 39.1°F, maximum density, and its volume is 0.1337 cu ft, therefore, the density of water is 8.345 lbs/0.1337 cu ft = 62.4 lbs/cu ft.

Specific gravity of a substance is the ratio of the density of the substance to that of water. This quantity is a pure numeric and indicates how many times the substance is as "heavy" (dense) as water. If a substance has a specific gravity of 2, then one cubic foot of that substance has a density of 2 × 62.4 = 124.8 lbs/cu ft.

Table D.1 Some Typical Specific Gravity Values

SUBSTANCE	SPECIFIC GRAVITY
Alcohol, ethyl	0.79
Ethylene glycol	1.115
Ethylene glycol (50%)	1.067
Water (pure)	1.00 (4°C)
Water (sea)	1.025 (15°C)
Air (for gases)	1.0

The thermal capacity of a substance is the number of British thermal units (Btus) needed to raise one pound of a substance through one degree Fahrenheit. Thermal capacity of substances varies somewhat with its temperature. Thermal capacity, commonly known as *specific heat* (C) of a substance, is expressed in Btu per lb-°F.

Table D.2 Some Typical Specific Heat Values

SUBSTANCE	SPECIFIC HEAT
Alcohol, ethyl	0.59
Ethylene glycol	0.57
Ethylene glycol (50%)	0.70
Water (pure)	1.00
Water (sea)	0.95
Air	0.24

One of the more common formulas used in HVAC calculations to determine the heat transferred in a system is:

$$H \text{ (Btu/hr)} = 500 \times GPM \times TD$$

this is derived from:

$$H \text{ (Btu)} = C \times W \times TD$$

The constant 500 is obtained by multiplying the weight of water in pounds per gallon, (8.335 lbs/gal) by 60 minutes per hour. This constant is correct if the weight of water is taken at 60°F. It is not true at all temperatures, for example, at 210°F water weighs 8.00 lbs/gal. Here the constant is 8.00 × 60 = 480. The value 500 is a recognized constant and is used in practical application problems. Steam tables provide density and weight for all temperatures for use where a more precise solution is needed. Properties of water are listed below for a few select temperatures:

Table D.3 Physical Properties of Water

TEMPERATURE OF WATER (°F)	DENSITY (lb/cu ft)	WEIGHT lb/gal)
32	62.414	8.3436
60	62.371	8.3378
100	61.996	8.2877
150	61.188	8.1797
210	59.862	8.0024

Another common formula used in HVAC calculations determines the horsepower required to pump water:

$$\text{Water Horsepower} = Q \times H \times SG / 3960$$
$$\text{Water Horsepower} = Q \times P \times SG / 1714$$

where: Q = Quantity of water in gallons per minute
H = Total dynamic head in feet
P = Pressure in pounds per square inch
SG = Specific gravity (1 for water)

Brake horsepower takes into account the mechanical efficiency of the pump (PME) and is used to determine motor size:

$$\text{Brake horsepower} = Q \times H \times SG / 3960 \times PME$$
$$= \text{Water Horsepower} / PME$$
$$\text{Pump Efficiency} = \text{Water Horsepower} \times 100 / BHP$$

There are three so-called pump laws:
1. The quantity of water delivered varies directly with the pump speed.
2. The head developed by the pump varies as the square of the speed.
3. The horsepower required by a pump varies as the cube of the speed.
1. $Q_2/Q_1 = N_2/N_1$
2. $H_2/H_1 = (N_2/N_1)^2$
3. $HP_2/HP_1 = (N_2/N_1)^3$

Head on a pump is expressed in feet of water, whereas pressure drop in piping due to friction is expressed in pounds per square inch. These figures are interchangeable as follows:

$$\text{Pounds per Square Inch} = 0/433 \times \text{Feet of Water,}$$
and
$$\text{Feet of Water} = 2.31 \times \text{Pounds per Square Inch}$$

c. Refrigeration

A ton of ice melting in 24 hours will remove heat at the rate of 12,000 Btu/hr. In mechanical refrigeration, a refrigeration system that removes heat at the rate of 12,000 Btu/hr is said to have a capacity of one ton.

$$\text{One ton (refrigeration)} = 12,000 \text{ Btu/hour}$$
$$= 200 \text{ Btu/minute}$$

Heat transfer in a HVAC system:

$$\text{Tons} = \text{GPM} \times \text{TD} / 24 \text{ (for water)}$$
$$= \text{CFM} \times \text{TD} / 3750 \text{ (for air)}$$
$$\text{Total heat (Btu/hr)} = 500 \times \text{GPM} \times \text{TD} \text{ (for water)}$$
$$= 4.5 \times \text{CFM} \times \text{TD} \text{ (for air)}$$

Energy cost for operating refrigeration equipment:

$$C = 0.746 \times P \times T \times H \times S / E$$

where: C = Operating Cost of Refrigeration (Dollars)
P = Power consumption (Brake HP per Ton)
T = Rated Refrigeration Capacity (Tone)
H = Full Load Operating Hours
S = Cost of Power (Dollars per Kilowatt-hours)
E = Motor Efficiency, (Decimal)

Cooling tower sizes are generally listed in nominal tons of refrigeration based on circulating 3 GPM/ton through the condenser with the rated quantity of air entering at 95°F and leaving at 85°F, with a 78°F design entering wet bulb. Total heat rejected (THR) by a cooling tower in Btu/minute:

$$\text{Btu/min} = \text{GPM} \times 8.33 \times (\text{Ewt-Lwt})$$

where: Ewt = Entering water temperature on tower (°F)
Lwt = Leaving water temperature off tower (°F)

The *Coeficient of Performance* (COP) of a heat pump is a method of measuring its heat output (in Btu/hr), in proportion to its power consumption (in Watts):

$$\text{COP} = \text{Heating Effect (Btu)} / \text{Input} (3.415 \times \text{Watts})$$

d. Boilers

Boilers are generally rated in horsepower.

1 Boiler Horsepower = 33,472 Btu/hr
= 9.8 KW
= 34.5 Lbs of steam (at and from 212°F)
= 30.0 Lbs of Steam (60°F Feed Water, 10 PSIG)
= 29.6 Lbs of Steam (60°F Feed Water, 100 PSIG)

The heat required to change ice to water is called *Latent Heat of Fusion.* (144 Btu/lb) The heat required to change water to steam is called the *Latent Heat of Vaporization.* (970.3 Btu/lb at atmospheric pressure). The term *Latent Heat* only is commonly used and the balance of the phrase is implied. The same amount of heat reverse the process, i.e., water to ice (144 Btu/lb), and steam to water (970.3 Btu/lb). Steam at the same temperature as boiling water is called *saturated steam*. Steam at a temperature higher than that of boiling water is called *superheated steam*.
Steam to heat water in hot water convertor is determined as follows:

$$\text{Steam (lbs/hr)} = 500 \times \text{GPM} \times \text{TD} \times H_{lh}$$
$$\text{Steam (lbs/hr)} = 8.34 \times \text{GPH} \times \text{TD} \times H_{lh}$$

where: GPH = Gallons per Hour

H_{lh} = Latent heat of steam, Btu/lb, at design temperature.

Boiler efficiency:

Fuel-to steam efficiency = Output (Btu/hr) / Input (Btu/hr)

For No.2 oil, use—140,000 Btu/gal
For No.6 oil, use—150,000 Btu/gal
For electric, use—Kwhr × 3.415

One volume of water evaporated at atmospheric pressure and 212°F becomes 1606 volumes of steam, i. e., one cubic foot of water becomes 1606 cubic feet of steam.

Index